Lipid Peroxidation Research

Edited by Mahmoud Ahmed Mansour

Published in London, United Kingdom

IntechOpen

Supporting open minds since 2005

Lipid Peroxidation Research
http://dx.doi.org/10.5772/intechopen.73713
Edited by Mahmoud Ahmed Mansour

Contributors
Bratko Filipič, Klemen Rihar, Adriana Pereyra, Hrvoje Mazija, Damir Đermić, Lidija Gradisnik, Arunachalam Muthuraman, Narahari Rishitha, Nallupillai Paramakrishnan, Bhaskaran Mahendran, Muthusamy Ramesh, Ana Lúcia Fernandes Pereira, Virgínia Kelly Gonçalves Abreu, Marija Vavlukis, Ana Vavlukis, Abubakar Babando Aliyu, Joseph Adesina Adewuyi, Patience Gangas, Hamisu Ibrahim, Adebayo Ojo Oyewale, Jonathan Ilemona Achika, Bukem Bilen, Mahmoud Ahmed Mansour

First published in London, United Kingdom, 2020 by IntechOpen
IntechOpen is the global imprint of INTECHOPEN LIMITED, registered in England and Wales, registration number: 11086078, 7th floor, 10 Lower Thames Street, London, EC3R 6AF, United Kingdom
Printed in Croatia

British Library Cataloguing-in-Publication Data
A catalogue record for this book is available from the British Library

Additional hard and PDF copies can be obtained from orders@intechopen.com

Lipid Peroxidation Research
Edited by Mahmoud Ahmed Mansour
p. cm.
Print ISBN 978-1-83968-547-7
Online ISBN 978-1-83968-548-4
eBook (PDF) ISBN 978-1-83968-549-1

For EU product safety concerns:
IN TECH d.o.o., Prolaz Marije Krucifikse Kozulić 3, 51000 Rijeka, Croatia, info@intechopen.com or visit our website at intechopen.com.

Meet the editor

Professor Mahmoud Mansour is currently a Professor of biochemistry at the College of Pharmacy, King Saud bin Abdulaziz University for Health Sciences. He received his pharmacy degree from the Al-Azhar University Cairo, Egypt in 1984, and his PhD degree in clinical biochemistry from the Karolinska Institute, Stockholm Sweden in 1992. From 1996 to 2016, he joined the King Saud University. As a fellow, he has contributed significantly to defining some aspects of the molecular biology of tissue injury and induction of liver and renal toxicity, with particular focus on the role of antioxidants in the prevention and treatment of different toxic effects. In addition, he has contributed notably to developing a new strategy for tumors especially in the liver. His research has shown that the proteasome inhibitor plays an important role in preventing carcinogenesis of the liver by metastatic tumors. He then moved to King Saud bin Abdulaziz University for Health Science, where he continued to work on the pathogenesis of cancer and on issues concerning the cellular and molecular mechanisms of cancer biology and the metastatic process. He is now conducting studies on liver cancer using gankyrin.

Contents

Preface

Lipid peroxidation is the major molecular mechanism that induces oxidative damage to cell structures and is also involved in the toxicity process that leads to cell death. Lipid peroxidation was initially studied as a mechanism for the damage to alimentary oils and fats; however, others thought that lipid peroxidation was the consequence of toxic metabolites (e.g. CCl_4) liberating a highly reactive species, the carbon trichloride methyl radicals (CCl_3). This radical quickly adds molecular oxygen to form the trichloromethylperoxy radical. Removal of hydrogen atoms from unsaturated fatty acids by such radicals creates carbon-centered lipid radicals. These radicals quickly add molecular oxygen to form lipid peroxyl radicals, thereby, initiating the process of lipid peroxidation. Unless scavenged by radical scavengers, these lipid peroxy-radicals in turn abstract hydrogen atoms from other lipids molecules, thereby initiating the process of lipid peroxidation.

Lipid peroxidation is a chain reaction initiated by the hydrogen abstraction or addition of an oxygen radical, resulting in the oxidative damage of polyunsaturated fatty acids (PUFA). PUFAs are more sensitive than saturated fatty acids because of the presence of a double bond adjacent to a methylene group that makes the methylene C-H bond weaker and therefore the hydrogen is more susceptible to abstraction. This leaves an unpaired electron on the carbon, forming a carbon-centered radical, which is stabilized by a molecular rearrangement of the double bonds to form a conjugated diene, which then combines with oxygen to form a peroxy-radical.

In pathological situations the reactive oxygen and nitrogen species are generated at higher than normal rates, and as a consequence, lipid peroxidation occurs with deficiency of endogenous antioxidants as alpha-tocopherol deficiency or reduced glutathione. In addition, in the presence of high concentrations of PUFAs and transition metals, biological membranes of cells and organelles are constantly being subjected to various types of damage. The mechanism of biological damage and the toxicity of these reactive species on biological systems are currently explained by the sequential stages of reversible oxidative stress and irreversible oxidative damage. Oxidative stress is understood as an imbalance between increased oxidants or decreased antioxidants. The concept implies the recognition of the physiological production of oxidants (oxidizing free radicals and related species) and the existence of operative antioxidant defenses. The imbalance concept recognizes the physiological effectiveness of the antioxidant defenses in maintaining both oxidative stress and cellular damage at a minimum level in physiological conditions.

Collectively, the book is a systematic and comprehensive review of lipid peroxidation with special emphasis on their role in different diseases.

Mahmoud Mansour
Professor,
King Saud Bin Abdulaziz University for Health Sciences,
Saudi Arabia

Chapter 1

Introductory Chapter: Free Radicals and Lipid Peroxidation

Mahmoud Ahmed Mansour

1. Introduction

During cellular metabolism, a potentially dangerous by-product named free radicals is liberated. They have several effects on cell survival, growth, and development and have remarkable effects in the pathogenesis of atherosclerosis, aging, development of cancer, and several other conditions including inflammatory diseases [1]. A free radical is characterized by containing in its outer orbit an unpaired electron [2]. During the process of adenosine triphosphate (ATP) production in the mitochondria, free radicals are generated by aerobic organisms. During the electron- transport steps of ATP production, due to the leakage of electrons from mitochondria, reactive oxygen species (ROS), superoxide anion (**$O_2^{-\bullet}$**) and hydroxyl (**$OH^{\bullet}$**) radicals, are generated. These free radicals through chemical reactions can lead to the production of hydrogen peroxide (**H_2O_2**). Based on the presence of Fe^{2+} ions, hydroxyl radicals are produced [3].

Free radicals are involved in several beneficial and harmful actions. Free radicals are involved in the signal transduction pathways that regulate cell growth [4] and reduction-oxidation (redox) status [3] and have a vital role in the defense polymorph nuclear leukocytes against infections as it acts as the first line of defense [5]. However, free radicals in excessive amounts can induce lethal chain reactions, leading to inhibition and inactivation of vital enzymes and many other proteins which are important subcellular elements needed for cell survival and leading to apoptosis [6]. Thus, functionally free radicals are considered a double-edged sword (**Figure 1**).

Reactive oxygen species include radicals such as superoxide (**$O_2^{-\bullet}$**), hydroxyl radical (**$HO^{\bullet}$**), nitric oxide (**$^{\bullet}NO$**), and non-radical species such as hydrogen peroxide (**H_2O_2**) and peroxynitrite (**$ONOO^-$**) [7].

Reactive oxygen species is produced both enzymatically and nonenzymatically. Enzymatic sources include NADPH oxidase located on the cell membrane of polymorphonuclear cells, macrophages and endothelial cells [8], and cytochrome P_{450}-dependent oxygenases [9]. Irreversible conversion of xanthine dehydrogenase to xanthine oxidase by mitochondrial protease uses molecular oxygen as electron acceptor and produces remarkable amounts of both $O_2^{-\bullet}$ and H_2O_2. Therefore it can provide another enzymatic source of both free radicals and also constitutes a source of $OH^{\bullet}$. The production of $O_2^{-\bullet}$ occurs nonenzymatically too via transfer of a single electrons to oxygen reduced coenzymes or prosthetic groups (e.g., flavins or iron sulfur clusters). Furthermore, previously reduced xenobiotics by certain enzymes (e.g., the anticancer agent Adriamycin or the herbicide paraquat) can also produce free radicals.

IntechOpen

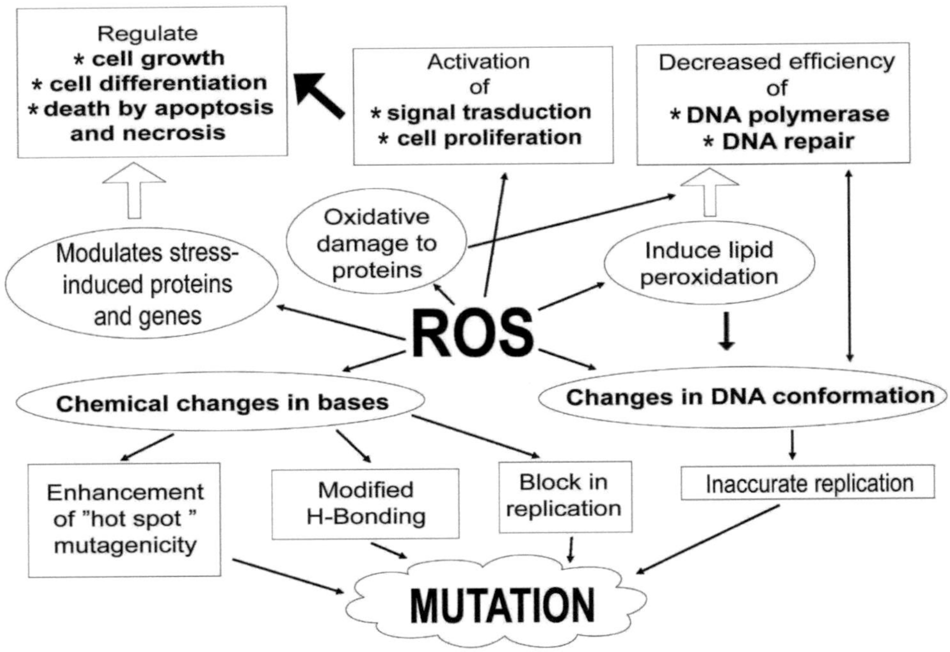

Figure 1.
ROS, oxidative damage, and human diseases. Interrelationship between the effect of imbalance in the reactive oxygen species (ROS) and their consequences on the cellular growth and the cellular function leads to DNA damage and mutation.

2. Role of oxidative stress in different diseases

The oxidative stress plays a role in the pathogenesis of different clinical conditions. Cancer diseases, diabetes mellitus, atherosclerosis, chronic inflammatory diseases, human immunodeficiency virus (HIV) infection (AIDS), ischemia–reperfusion injury, and sleep apnea are important examples. The previously mentioned disease can be classified into two categories [10]. In the first category, a pro-oxidative shift in the systemic thiol/disulfide redox state is in parallel with impaired glucose clearance, suggesting that the mitochondria of the skeletal muscle may be the primary site of elevated ROS production; these conditions may be referred to as "mitochondrial oxidative stress" which is clearly in diabetes mellitus and cancer [11]. The second category is based on excessive stimulation of NADPH oxidase activity by cytokines or other agents and therefore refers to "inflammatory oxidative conditions." In this case, elevated free radical levels or alteration of intracellular glutathione levels is often associated with pathological changes indicative of a dysregulation of signal transduction and/or gene expression, represented by a change in the expression of cell adhesion molecules [12].

3. Lipid peroxidation and incidence of cancers

There is clear evidence supporting the role of lipid peroxidation in the induction of selected human cancers, including the kidney, liver, and skin. Estrogen treatment induces lipid peroxidation and subsequently increases the incidence of renal cell cancer in experimental models [13, 14]. Based on this mechanism, it has been hypothesized that estrogen increases breast cancer risk as lipid peroxidation may be one mechanism [15]. But estrogen induces renal cancer or liver cancer in this experimental model, not breast cancer.

In contrast, there is evidence favoring lipid peroxidation as an anticarcinogenic mechanism in breast cancer. It has been confirmed that higher level of lipid peroxidation is usually associated with lower rate of cell proliferation. Therefore, there is an inverse relationship between the concentrations of lipid peroxides and the rate of the cell proliferation [16]. This is supported by the observation that tumor cells are more resistant to lipid peroxidation than normal cells [17]; indeed, it was shown that in hepatomas, the higher the growth rate of the tumor, the lower the microsomal phospholipid content and the degree of fatty acid unsaturation [16]. Hosmark and Lystad [18] have also reported that low levels of polyunsaturated fatty acids and cytochrome P_{450} and elevated levels of lipid-soluble antioxidant alpha-tocopherol in the hepatoma cells are the main causes behind lower rate of lipid peroxidation.

It has been reported that lipid peroxidation represents a protective mechanism in breast cancer. Decreased plasma malondialdehyde (MDA), which is a marker for lipid peroxidation, has been significantly associated with severity of prognosis factors for breast cancer. A significant lower plasma level of MDA was detected in patients with large tumors or in whom nodes and/or metastasis was present [19, 20].

Author details

Mahmoud Ahmed Mansour
Department of Pharmaceutical Sciences, Pharmacy College, King Saud Bin Abdulaziz University for Health Sciences, Riyadh, Saudi Arabia

*Address all correspondence to: mansoura@ksau-hs.edu.sa

References

[1] Milisav I, Ribarič S, Poljsak B. Antioxidant vitamins and ageing. Sub-Cellular Biochemistry. 2018;**90**:1-23

[2] Boveris A. Biochemistry of free radicals: From electrons to tissues. Medicina (B Aires). 1998;**58**(4):350-356

[3] Cleveland JL, Kastan MB. Cancer. A radical approach to treatment. Nature. 2000;**407**(6802):309-311

[4] Shukla S, Mishra R. Level of hydrogen peroxide affects expression and sub-cellular localization of Pax6. Molecular Biology Reports. 2018;**45**(4):533-540

[5] Gizinger OA, Korkmazov AM, Sumerkina VA. Functional activity of neutrophils and local manifestations of oxidative stress in the mucous membrane of the nasal cavity in the early postoperative period. Vestnik Otorinolaringologii. 2019;**84**(2):40-45

[6] Singh A, Kukreti R, Saso L, Kukreti S. Oxidative stress: A key modulator in neurodegenerative diseases. Molecules. 2019;**24**(8)

[7] Elsner J, Kapp A. Reactive oxygen release. Methods in Molecular Biology. 2000;**138**:153-156

[8] Jones OT. The regulation of superoxide production by the NADPH oxidase of neutrophils and other mammalian cells. BioEssays. 1994;**16**(12):919-923

[9] Cassagnes LE, Perio P, Ferry G, Moulharat N, Antoine M, Gayon R, et al. In cellulo monitoring of quinone reductase activity and reactive oxygen species production during the redox cycling of 1,2 and 1,4 quinones. Free Radical Biology & Medicine. 2015;**89**:126-134

[10] Ramos-Tovar E, Muriel P. Free radicals, antioxidants, nuclear factor-E2-related factor-2 and liver damage. Journal of Applied Toxicology. 2019

[11] Białas AJ, Sitarek P, Miłkowska-Dymanowska J, Piotrowski WJ, Górski P. The role of mitochondria and oxidative/antioxidative imbalance in pathobiology of chronic obstructive pulmonary disease. Oxidative Medicine and Cellular Longevity. 2016;**2016**:7808576

[12] Lenz AG, Karg E, Brendel E, Hinze-Heyn H, Maier KL, Eickelberg O, et al. Inflammatory and oxidative stress responses of an alveolar epithelial cell line to airborne zinc oxide nanoparticles at the air-liquid interface: A comparison with conventional, submerged cell-culture conditions. BioMed Research International. 2013;**2013**:652632

[13] Gago-Dominguez M, Castelao JE, Yuan JM, Ross RK, Yu MC. Lipid peroxidation: A novel and unifying concept of the etiology of renal cell carcinoma (United States). Cancer Causes & Control. 2002;**13**:287-293

[14] Gago-Dominguez M, Castelao JE. Lipid peroxidation and renal cell carcinoma: Further supportive evidence and new mechanistic insights. Free Radical Biology & Medicine. 2006;**40**:721-733

[15] Gago-Dominguez M, Castelao JE, Pike MC, Sevanian A, Haile RW. Role of lipid peroxidation in the epidemiology and prevention of breast cancer. Cancer Epidemiology, Biomarkers & Prevention. 2005;**14**:2829-2839

[16] Chajes V, Sattler W, Stranzl A, Kostner GM. Influence of n-3 fatty acids on the growth of human breast cancer cells in vitro: Relationship to peroxides and vitamin-E. Breast Cancer Research and Treatment. 1995;**34**:199-212

[17] Gago-Dominguez M, Castelao JE, Sun CL, Van Den Berg D, Koh WP, Lee HP, et al. Marine n-3 fatty acid intake, glutathione S-transferase polymorphisms and breast cancer risk in postmenopausal Chinese women in Singapore. Carcinogenesis. 2004;**25**:2143-2147

[18] Høstmark AT, Lystad E. Growth inhibition of human hepatoma cells (HepG2) by polyunsaturated fatty acids. Protection by albumin and vitamin E. Acta Physiologica Scandinavica. 1992;**144**(1):83-88

[19] Saintot M, Astre C, Pujol H, Gerber M. Tumor progression and oxidant-antioxidant status. Carcinogenesis. 1996;**17**:1267-1271

[20] Gerber M, Astre C, Segala C, Saintot M, Scali J, Simony-Lafontaine J, et al. Tumor progression and oxidant-antioxidant status. Cancer Letters. 1997;**114**:211-214

Chapter 2

Role of Lipid Peroxidation Process in Neurodegenerative Disorders

Arunachalam Muthuraman, Narahari Rishitha, Nallupillai Paramakrishnan, Bhaskaran Mahendran and Muthusamy Ramesh

Abstract

Lipid peroxidation is one of the primary events of the cell injury process. In pathophysiological condition, it is undergoing the initiation of organ damage. Various free radicals are playing a key role in this lipid peroxidation process. Free radical associated organ damage involves the three major phases, that is, initiation, propagation and termination. The primary source of various free radical formations is mediated through the pathophysiological function of mitochondria. Lipid peroxidation is contributed to the multiple neurodegenerative disorders. Thus, the various endogenous cellular anti-oxidant systems are regulated lipid peroxidation process and control the neurodegenerative action. Some of the molecules are targeted to attenuate the lipid peroxidation and their mediators for the prevention of neurodegeneration.

Keywords: malondialdehyde, vascular dementia, Alzheimer disease, multiple sclerosis

1. Introduction

Lipid peroxidation (LPO) is a complex process of the cellular system. The reactive oxygen species (ROS) such as superoxide anion, hydroxyl radical and hydrogen peroxide radicals play a key role in the process of lipid peroxidation [1]. Oxygen radicals, that is, superoxide anions are a primary bioactive molecule in LPO process. LPO process releases the various metabolic intermediated products such as malondialdehyde (MDA) and 4-hydroxynonenal (HNE) [2]. Further, the accumulation of LPO products; even it is small amounts induces the cell death process via multiple cell signaling process [3, 4]. ROS are ready to attack/interact with various bio molecules like polyunsaturated fatty acids (PUFA) of the fatty acid membrane. Then, LPO products are initiating the multiple self-propagative chain reactions [5]. This LPO associated destruction of membrane lipids and their intermediate/end-products are potentially dangerous for the various cells, tissues and organs. However, the LPO process is overcome by the cellular enzymatic system such as catalase (CAT) & superoxide dismutase (SOD); and non-enzymatic molecules like vitamins A and E [6]. Thus, lipid peroxidation is a self-propagating chain-reaction and involves the multiple ways of LPO product formation; so the availability of few lipid products can cause the significant tissue damage [7]. However, the extensive research about the lipid peroxidation process and products are not yet been studied. The precise action and regulatory function of LPO in various pathophysiological conditions of neurodegenerative disorders such as Alzheimer disease (AD), dementia, Parkinson disease (PD), Huntington disease (HD), multiple sclerosis (MS), amylotropic lateral sclerosis (ALS), stroke and neuropathy are needed to be investigated [8, 9]. This book chapter is focused on the role of the lipid peroxidation process in neurodegenerative disorders.

2. Relationships of free radicals and LPO

LPO occurs due to the generation of free radicals process in the biological system. The free radicals are readily attacking the phospholipids of cell membrane leads to degrading the lipids action [10]. This rapid reaction towards lipid membrane is due to the availability of multiple double bonds between methylene ($-CH_2-$) bridges and reactive hydrogen atoms. This is most frequently occurs with polyunsaturated fatty acids (PUFA) [11]. The free radical reactions are self-perpetuating chain reactions and it is highly reactive molecules with a various biomolecule such as protein, lipids, mitochondria, endoplasmic reticulum and DNA due to its unpaired electrons [12]. In addition, these radicalsare existed very short duration, that is, 10^{-9}–10^{-12} second. However, before diverging of these free radicals are reacts with another molecule; to make their own stability by attracting or donate the electrons [13]. Generally, controlled generation of free radicals under normal body condition is good for the physiological process; because it enhances the immune cell activation and stimulates the various cellular systems.

Unfortunately, the large quantity of free radical generation causes the cell death. The abundant generation of free radicals is occurred due to abnormal metabolic and homeostatic functions. Normally the raising of free radicals is controlled by various endogenous anti-oxidant defense systems. There are two major factors are employing the activation of free radical associated lipid peroxidation process. The primary factor is increasing of cytosolic free radical concentration and it occurs due to enzymatic and mitochondrial mediated reactions [14]. Another factor is the reduction anti-oxidant fence system and it occurs due to the changes in the normal endogenous enzymatic pool. The imbalanced action of free radicals ate responsible for the LPO. LPO process not only occurs in cell membrane lipids; it also occurs in mitochondria, endoplasmic reticulum and nuclear membrane. This intracellular reaction of LPO is enhancing the cell death process [15]. Later stage of LPO process is also enhanced release of various biomolecules which activates the paracrine actions. The overall effect is procuring the organ and system failure. Here some of the examples are listed for the lipid peroxidation associated disorders such as cancer, atherosclerosis, myocardial infarction, coronary artery disease, liver failure, renal failure and autoimmune disease. In addition, the numerous reports are documented that, LPO process also enhances the pathogenesis of neurological disorders such as Alzheimer's disorder; Parkinson disease; multiple sclerosis, vascular dementia, stroke and neuropathic pain [16].

3. Molecular mechanism of free radicals, antioxidant and toxicity reactions

The free radical (R^*) formation is mainly due to the abnormal bio-activation process sometimes due to xenobiotic reactions such as cytochrome P_{450}; prostaglandin synthase and lipoxygenase reactions. Free radical associated lipid peroxidation reaction has three stages. (1) *Initiation stage*: in this stage, free radicals are attacking the covalently bonded molecules due to its high affinity [17]. The covalent bond containing bio-molecules are DNA, proteins, lipids and phospholipids. The membrane phospholipids, that is, PUFA is a primary target in the cellular system. In this stage fatty acid radical, that is, lipid radical (L*) is produced. (2) *Propagation stage*: the primary noticeable initiators of propagation stage of free radicals are reactive oxygen species (ROS; i.e., OH• and HOO•). This stage another lipid radicals, that is, lipid peroxy radical (LOO*) and lipid hydroperoxide (LOOH) are produced. Here, LOOH is non-radical. Whereas, when radicalsreact with non-radical molecules; non-radical molecules ready to changes their own property lead to form another active radical. This process is called as "chain reaction mechanism". (3) *Termination stage*: in this stage; different molecules are involved to speed up termination free radical—LPO reaction by neutralizing free radicals such agent is also called antioxidants such as vitamin E and vitamin C. Some of the anti-oxidants molecules are presents within the body like superoxide dismutase, catalase and glutathione peroxidase. These molecules are actively converting the LOO* and LOOH molecules to stable lipid alcohol (LOH). It is non-toxic to the biological system. Generally, the radical reaction stops when two radicals are reacted and produces the non-radical species. However, these reactionsare happening, when the concentration of radical species is higher. In pathological conditions, the failure of the termination stage of LPO process leads to produce the multiple lipid peroxidative products [18–20]. The free radical (R*) associated lipid peroxidation in two different phases. The first phase is the lipid phase; it occurs in the cell membrane. The second phase is an aqueous phase; it occurs in the cytosolic region of the cell. Antioxidants are regulated inboth phases of the LPO process [21].

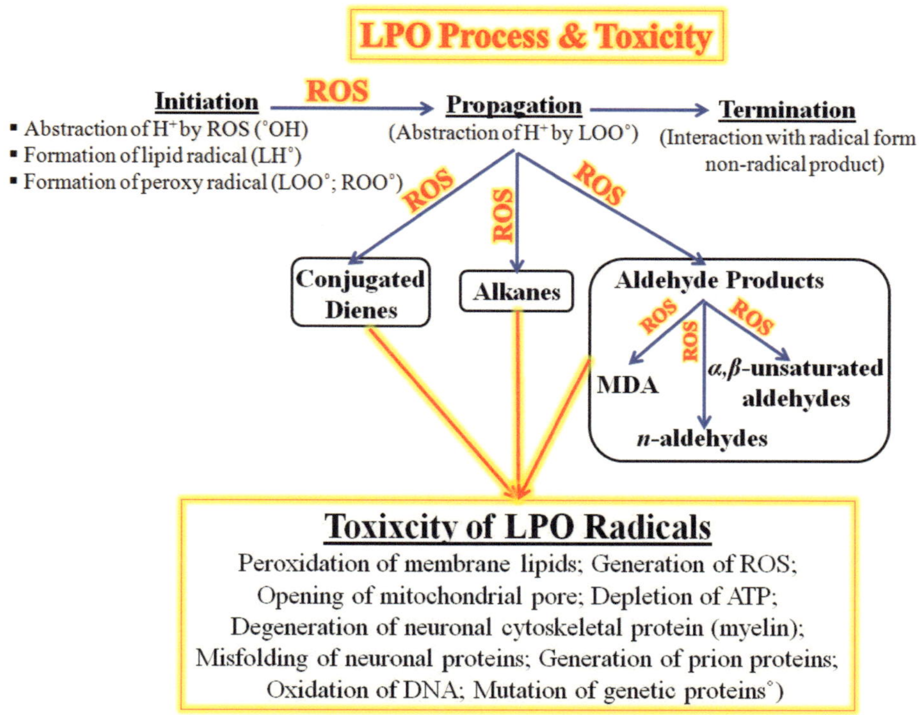

Figure 1.
This illustration is showing the reactions between free radicals, lipids and antioxidant molecules. Mechanism of LPO process and lipid radical associated neurotoxicity are undergoes the following actions, that is, initiation, propagation and termination. Whereas, the propagation stage of lipid radicals are induce the formation of various lipid radicals. Finally, it interacts with multiple levels of biomolecular changes leads to cause the neurodegeneration.

In lipid phase, PUFA methylene (—H—) bridges are converts to PUFA oxy (—OO*—) radical. An antioxidant such as tocopherol (Toc-OH) induces the PUFA oxy (—OO*—) radical conversion to PUFA oxy (—OOH—) molecule. PUFA peroxy (—OOH—) bridged molecule biologically non-toxic. Further nontoxic PUFA peroxy (—OOH—) bridged molecules are converted to PUFA peroxy (—OOH*—) molecule with the action of phospholipase A_2 (PLA_2). In this process is shifting the lipid phase to the aqueous phase of the reaction [22]. The PUFA peroxy (—OOH*—) molecules are toxic to the cellular system [23]. These molecules are regulated by two endogenous enzymatic anti-oxidant systems such as catalase and glutathione peroxidase. Catalase is directly converting the PUFA peroxy (—OOH*—) molecules to PUFA peroxy (—OH—) molecules. Glutathione peroxidase enhances the neutralizing of PUFA peroxy (—H_2O_2—) molecules with the help of selenium (Se). In the cytosol, the superoxide accumulation occurs due to the ETC reaction and other mitochondrial reactions. The aqueous phase superoxide ions are neutralized by superoxide dismutase enzyme. The failure of free radical—LPO reactions leads to activates cell death; proximal and distal tissue damage; and organ failure process [24]. The interaction of free radical, lipid peroxidation and toxicity mechanism is showed in **Figure 1**.

4. By-products of lipid peroxidation

The LPO of unsaturated lipids and ROS attack on unsaturated lipids are producing the variety of oxidation products. The primary oxidation products of lipid are lipid hydroperoxides (LOOH). In addition, The LPO process releases the additional by-products via metabolic conversion of lipids and their radicals. By-products of LPO are reactive aldehydes like malondialdehyde (MDA), 4-hydroxynonenal (HNE) propanal and hexanal. MDA is identified as the potent mutagenic product of lipid peroxidation; whereas, another LPO products, that is, HNE is toxic to the cellular system but not mutagenic. LPO products are categorized into two ways; first one is primary LPO product, that is, LOOH and their adduct products. Another one is the secondary LPO product, that is, MDA and their adducts [25]. Primary lipid peroxidation products (hydroperoxides) are formed at the propagation phase of the LPO process. The hydroperoxide group can be attaching to multiple lipid molecules such as free fatty acids; triacyl glycerols; phospholipids; and sterols. LOOH are more stable products and found in serum. Hence, the LPO products can be detectable for identification of cellular oxidative stress. The LOOH are targeted to various reduction reactions; which lead to either inhibition of peroxidative damage and/or induction of peroxidative damage [3]. The inhibition of peroxidative damage is due to decomposing of hydroperoxides, namely two-electron reduction [26]. Some of the enzymes are responsible for this two-electron reduction process of hydroperoxides. Those enzymes are selenium-dependent glutathione peroxidases (GPx) and selenoprotein P (SeP). The induction of peroxidative damage is due to decomposing of hydroperoxides, namely one-electron reduction in the initiation and/or propagation steps of LPO process [27]. These conditionsenhance the formation of new LOOH, that is, lipid peroxyl radical and alkoxyl radicals by redox cycling process [28].

A secondary lipid peroxidation product (MDA) is a by-product of arachidonic acid and PUFAs decomposition. This reaction occurs by enzymatic or non-enzymatic reaction process. MDA has a dose-dependent dual role function; it is chemically stable and membrane-permeable compare to ROS; whereas, MDA is less toxic than HNE and methylglyoxal (MG) products. The enzymatic production of MDA is occurred during the biosynthesis of thromboxane A_2 (TAX-A_2) from arachidonic acid metabolism by the action of TAX-A_2 synthase [29]. The non-enzymatic production of MDA occurs with a mixture of lipid hydroperoxides during lipid peroxidation process. The non-enzymatic production of MDA is occurring under specific conditions. The metabolism of MDA involves via mitochondrial aldehyde dehydrogenase actions; which leads to form decarboxylated products, that is, acetaldehyde; and it is further oxidized by aldehyde dehydrogenase to form acetate; carbon-di-oxide (CO_2); and water (H_2O) [3].

Both primary and secondary by-products of LPO are playeda key role in the various pathophysiological process including neurodegenerative disorders. MDA and HNE products are interfering with the various biological processes such as genetic; physiological; and pathological to control the intrinsic and extrinsic factor influences [30]. In genetic aspects, LPO can be controlled by diet; environment; and habits. Various LPO studies are carried out in experimental animal models. The following agents are reported to produce the LPO in the biological system. Cumene hydroperoxide, it is used in the chemical and pharmaceutical industry as a catalytic agent [31]. Tert butyl hydroperoxide is an organic oxidizing agent and acts as a bleaching agent; and initiator of polymerization. Carbon tetrachloride (CCl_4) is a toxic carcinogenic agent and it is used as a solventfor the industrial degreasing operations. In addition, it is also used as pesticides and chemical intermediate of refrigerants. Quinolinic acid (QA) is produces the potent neurotoxic metabolite of kynurenine metabolism intermediates and involves in the pathogenesis of neurological diseases [32]. Furthermore, some of the transition metals ions also play a role for LPO process due to its pro-oxidant effect. The major transition metals for the LPO is copper; chromium; cadmium; nickel; vanadium; manganese; and iron [3, 33–35].

In aging condition; iron plays a key role in the neurological disorders. Iron accumulation in CNS causing the lipid peroxidation process and it causes the motor dysfunction and other neurological disorders [36]. Now, iron dependent cell death is linked with the lipid peroxidation process. This phenomenon also called as "*ferroptosis*". Ferroptosis is a process and produces the non-apoptotic manner of cell death regulation process [37]. This iron dependent regulation of cell death is characterized by the accumulation of LPO products. Moreover, ferroptosis process can be controlled by the lipid acting antioxidants and potent iron chelators [37, 38]. Some of the molecules are identified as ferroptosis regulators such as erastin and rat sarcoma selective lethal 3 (RSL3) proteins [39].

5. Hazards action of lipid peroxidation products

The developed LPO products must be degraded in the biological system by enzymatic and non-enzymatic manner. If the termination of oxidized LPO products is not enough; it's causing the potential tissue damage to the cell membrane. The first phase of LPO is altered by the cytosolic enzymatic and calcium pool [40]. The second phase of LPO is shown to activation of macrophage followed by oxidation of PUFA [41, 42]. In addition, LPO products also interact with multiple catalysts and heavy metals such as iron and copper metals. A ferrousform of iron is highly interactive metals with lipid peroxidative products and it also documented to produce the potent neurodegenerative process [33, 42].

The summary of LPO induced neurodegeneration is due to free radical associated activation of membrane lipid oxidation and subsequent alteration of free radical, mineral, metal and enzymatic pool activation in the cytosol of every cellular system including nervous tissue [42, 43]. In a neuron, these actions are rapid and specialized targeted proteins, that is, cytoskeleton, misfolding and prion proteins [44]. Some of the actions occur in mitochondria and nuclear levels [45]. This multiple molecular and cellular actions are responsible for the sustained activation of

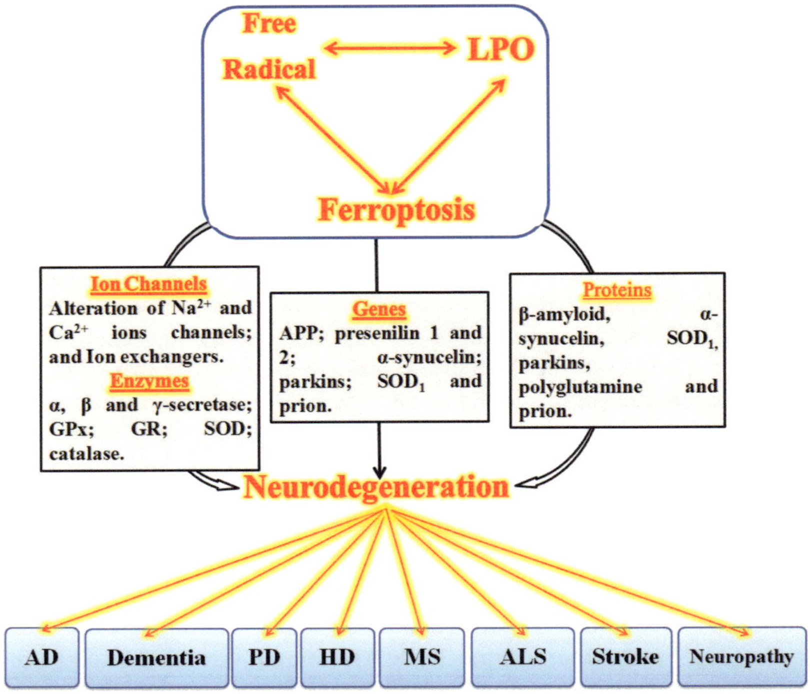

Figure 2.
Mechanism of LPO associated neurodegenerative disorders. The initial event of free radicals enhances the lipid peroxidation of products, which leads to alter the ion channels; enzymes; genes and proteins. The net effects of LPO products are cause the Alzheimer disease (AD); dementia; Parkinson's disease (PD); Huntington disease (HD); multiple sclerosis (MS); amylotropic lateral sclerosis (ALS); stroke; and neuropathy.

the neurodegenerative process leads to produce the multiple neurological disorders with neurodegeneration [46, 47]. This hazards reaction of LPO on targeted biomolecule for neurodegenerative disorders are illustrated in **Figure 2**. The following section is explaining about the LPO role in neurodegenerative disorders like Alzheimer disease (AD), Parkinson disease (PD), multiple sclerosis (MS), amylotropic lateral sclerosis (ALS), stroke and neuropathy.

6. Role of LPO in Alzheimer disease

LPO is one of the primary key factors for the progress of neurodegenerative disorders. The pathogenesis of AD is also documented that, oxidative stress enhance the progress of neurodegeneration via free radical associated LPO reactions [48]. The progress of AD rate is higher in developing countries and it causes the reduction of quality of life [49]. The various etiological factors are identified in the development of AD such as smoking, alcohol, hypertension, atherosclerosis, diabetes, hypercholesterolemia, aging and trauma. In addition, specialized neuronal proteins such as amyloid precursor protein (APP), the Aβ peptide, neurofibrillary tangles, Tau proteins, and amyloid plaques are plays a role in the alteration of neuronal function and its enhance the neurodegeneration and development of AD [50, 51]. Furthermore, these changes are due to alteration of multiple cellular signaling cascades such as over activation of α, β and γ secretase; beta-site APP-cleaving enzyme 1 and 2 ($BACE_1$ & $BACE_2$); presenilins proteins 1 and 2 (P_1 & P_2); apolipoprotein E-epsilon4 variant (ApoE-ε4); $alpha_2$-macroglobulin; and calcium dependent kinases [49, 52–55]. LPO and their metabolites contribute to the development of neurovascular complication and neurodegeneration by activation of the above signaling pathways. Recently, ferrous ion dependent free radical generation, lipid peroxidation and activation of amyloid proteins are responsible for the development of AD [56]. The classical medicines of nootropic agents like donepezil; rivastigmine; galantamine; and memantine are enhanced the memory function [57]. Some of the agents are anti-inflammatory agent such as zileuton is regulated in the neurodegenerative process via anti-lipidperoxidative and anti-inflammatory actions [58]. However, there are limited studies reported that anti-lipid peroxidative agents are ameliorated the AD [59, 60].

7. Role of LPO in dementia

Dementia is one of the leading causes of changes in quality of life and death. Now a days, various pathophysiological factors are employed in the progression of dementia especially VaD; such conditions are hypertension, diabetic mellitus, renal failure, cardiac failure and bone fracture [61–63]. The hallmark of dementia is still complicated and some of the reports revealed that α-synuclein protein alteration enhances the neurodegeneration and dementia symptoms [64]. In addition, the genetic predisposition is identified as causing factors for dementia disorders. Recently, some of the newer molecular proteins are explored in the neurodegeneration as well as dementia such proteins are α-synuclein, ubiquitin, parkin, neurofilaments and chaperone proteins. These all changes are observed in the moderate and chronic condition of dementia disorders. Whereas, the initial effects of the neurodegeneration process is to express the development of free radical and LPO associated oxidative stress environments in the nervous system [18]. However, the specific molecules for the LPO regulators are not studied in dementia disorders. Whereas, numerous studies revealed that, the memory enhancing drugs like donepezil; rivastigmine; galantamine; and mementine are reduced the lipid peroxidative products reaction. Hence, the direct and specific modulatory agents are required for the management of dementia disorders [65]. Further, dementia disorders are arising with multiple pathophysiological factors; so the multi-targeted drug approaches will be more effective to prevent the dementia disorders. Here, the LPO targeted drugs is one of the primary facts for the treatment of dementia disorders.

8. Role of LPO in Parkinson disease

Parkinson disease (PD) is one of the extrapyramidal disorders and it occurs due to dopaminergic neurodegeneration and alteration of dopamine levels in the brain. Symptomatically, it affects the motor neuron and produces the abnormalities of body movements. The LPO process induces the primary form of sporadic type of Parkinson disease (PD) [66]. Mainly, dopaminergic neuron damage occurs in the substantia nigra, locus ceruleus, dorsal motor nucleus and substantia innominata [67]. The sign of movement problems is festinating gait; rigidity of limbs; poverty of voluntary movement; and rolling type of tremor. The progressiveness of PD is very slow [68]. The various types of cells are employed in the dopaminergic neurodegeneration such as neutrophils; macrophages; and astrocytosis [69]. This all process is altered by free radicals and lipid peroxidative reactions. LPO reactions affect multiple biomolecules such as such as membrane lipids, proteins (especially α-synculein); mitochondrial pore; endoplasmic reticulum; liposomes; peroxisomes; lysomes including DNA and RNA [51]. Therefore, the reduction and controlling LPO process is able to treat the PD. However, the limited agents are tested in the management of dopaminergic neurodegeneration. Hence, the more extensive studies are required for the treatment of PD with direct and specific anti-lipid peroxidative drugs or combination with subsequent cell signal arresting molecules [70, 71].

9. Role of LPO in Huntington disease

Huntington disease (HD) is one of the autosomal dominant diseases. And, it is also one of the neurodegenerative disorders. It occurs between the age of 20 and 50 years. The symptoms of HD are a chorei form of body movements; dystonia; the paucity of movement and progressive loss of neuron [45]. Genetically, it affects the 4th chromosomal gene; huntingtin (HTT) protein mutation and misfolding. The main area of the brain is affection in caudate; putamen nucleous; globus pallidus; and nucleus accumbens leads to cause the atrophy of brain via the neurodegenerative process [32, 72]. The microscopical observation revealed that small spiny neurons of the caudate and putamen cause the astrocytosis. The chronic event of cellular neurodegenerative process enhances the neuronal oxidative stress with an accumulation of free radicals and LPO process [73]. During the early stage of LPO process; neuronal cells losses the cell contents; shrinkage of caudate nucleolus; shrinkage and dilatation of the anterior horns of lateral ventricles [45, 72]. Moreover, LPO process enhances the alteration of various neurotransmitters such as a gamma aminobutyric acid (GABA); acetylcholine; and substance P [74, 75]. The various medicines are reported to treat the HD via inhibition of LPO process [76]. Therefore, LPO targeted medicines are useful for the management of HD and neurodegenerative disorders.

10. Role of LPO in multiple sclerosis

Multiple sclerosis (MS) is a known demyelinating disease and its causes the dysfunction of the central nervous system. Myelin produces the insulation coating on the nervous system like electrical wires. The pathogenesis of MS is shown by attacks of neuronal proteins of CNS such as myelin proteins. In addition, the LPO products are play key role in the pathogenesis of MS [77]. This myelin protein is essential for neuronal function because of its support of the neuronal signal conduction without loss of strength. Whereas, the loss of myelin proteins by immunological abnormalities are cause the neuronal dysfunction and symptoms of MS. Further, these neuronal damages are permanent and difficult to the MS [78]. The symptoms of MS are numbness or weakness in limbs, one side of your body, legs and trunk; partial or complete loss of vision and double vision; tingling and pain; tremor, lack of coordination, unsteady gait, slurred speech, fatigue, dizziness; and bowel and bladder function. These all symptoms are due to damage of myelin proteins of CNS [79]. The various factors are employed in the pathogenesis of MS such as genetic abnormalities; environmental factors (smoking and alcohol); and autoimmune disease (thyroid disease, type 1 diabetes and inflammatory bowel disease) [80]. The MS is a progressive demyelinating process, which is due to the over the action of the immunological system. The immunological proteins are inducing the multiple physiological processes. Whereas the over activity of immunological proteins causes the neurological damage via activation of the demyelinating process [81]. This chronic progressive is due to the free radical and LPO process. The limited drugs are used for the treatment of MS and it also documented that, reduce the LPO products accumulation. However, the LPO targeted drugs are not tested in MS disease. So, extensive studies are required LPO targeted medicine for the treatment of MS.

11. Role of LPO in amylotropic lateral sclerosis

Amyotrophic lateral sclerosis (ALS) is another neurodegenerative disorder of a nervous system and it affects the CNS nerve cells. In addition, it affects the neuronal transmission between the brain and the spinal cord. This leads to affects the neuromuscular function and produces the trouble for walking, running and writing; and speech problems [82]. In addition, it also produces the respiratory failure; it is a very serious complication of ALS. The primary target of neurological damage is upper motor neurons of cranial motor nuclei, and Betz cells of neocortex lead to produce the muscle atrophy and astrocytosis [83]. Sometimes ALS is also produced in the lateral column degeneration with gliosis; so it is called "sclerosis". The primary etiology of ALS is due to the mutations of superoxide dismutase-1 (SOD1) gene [78]. The SOD gene associated proteins are responsible for the reduction of cellular free radicals leads to reduce the LPO products associated with protein and DNA damage [74, 84]. Therefore, LPO targeted drug is essential for the treatment of ALS. However, there is no report is documented for the treatment of ALS.

12. Role of LPO in stroke

Stroke is one more neurodegeneration disorders and it occurs due to the cerebrovascular accidents. The primary symptoms of stroke are a cluster headache, motor impairment, paralysis and death [85–87]. The various risk factors are employed in the pathogenesis of strokes such as smoking, alcohol, obesity, trauma, hypertension, diabetes and renal failure [88, 89]. In addition, various physiological processes are altered with multiple ethiological factors such as hypoxia, ischemia, aneurysm and immune activation leads to platelet activation, thrombosis, immune cell activation decreasing of ATP content, mitochondrial damage, ionic imbalance and glial cell activation [90–92]. Furthermore, the chronic neurological damage is due to the accumulation of lipid peroxidative products leads to damage to the neurological system leads to produce the stroke symptom [93, 94]. The free radical and LPO are well documented to produce neurodegeneration in stroke condition [95]. However, the tested compounds are show in the alteration of LPO products in the neurological system. But, the direct and specific action on LPO products is not documented in stroke disorders. So, the extensive studies are required to manage the stroke disorders with direct and newer molecule for the reduction of LPO products [96].

13. Role of LPO in neuropathy

The pathogenesis of neuropathy is occurred due to the neurodegenerative disorders. In diabetic condition, it produces the multiple types like autonomic neuropathy, somatic neuropathy, peripheral and central neuropathy. The etiology of neuropathy is due to multiple factors such as ischemia, trauma, hypoxia, free radical and lipid peroxidation [9, 97]. These factors alter the cellular and molecular events [98, 99]. The anatomy of microvascular system is affecting by the above factors and it raising the oxidative stress environment along with DNA damage, ATP depletion and activation of ferroptosis [100]. The symptoms of neuropathy are the development of autotomy (self-amputation of figures); allodynia (triggering of a pain response from stimuli, but it does not provoke pain in normal condition); hyperalgesia (extreme reaction of painful stimuli); and numbness (unusual prickling sensation). The chronic condition of neuronal firing enhances the neurodegeneration and reduces the quality of life [101]. In diabetic condition, the specialized etiology involved in the pathogenesis of neuropathy; such factors are raising blood glucose, sorbitol, aldose, ketone bodies; and advanced glycation end products (AGE) [102]. This all molecules are interlinked with oxidative stress. The accumulation of free radicals and the LPO process involves in the all pathological situation of neuropathy and neurodegenerative process; which leads to producing the subsequent activation of ionic imbalance; vascular damage; and neuronal damage [103, 104]. The LPO products play a crucial role in the pathogenesis of neuropathy [105, 106]. However, the directly acting agents on LPO products and specific molecules based actions still need to investigate for the management of neuropathy [107].

14. Inhibiting of LPO products

The various approaches are attempted to prevent the LPO products formation. There are two approaches are employed in the prevention of LPO products such as (1) prevention of LPO products formation by both enzymatic and non-enzymatic manner; (2) elimination of LPO products [108, 109]. Prevention of LPO products formation is a primary successful method with anti-oxidants such as vitamin C and vitamin E are inhibits the LPO [110]. The alternative method of LPO inhibition is deuteration of PUFA at double bond of methylene bridges (bis-allylic sites) leads to reduce the chain reaction process [111]. The major deuteriated PUFAs is 11,11-D2-ethyl linoleate is identified as suppressing agent for LPO process at low level concentration [112]. The primary diagnostic methods for end-products of LPO, that is, malondialdehyde (MDA) is thiobarbituric acid reactive substances (TBARS) assay. Recently, some immunochemical method is employed for the detection of HNE-histidine adducts in biological tissue and fluid samples [113]. Based on this basic fundamental of LPO formation and diagnostic procedure; some of the molecules are identified for the prevention of LPO process and products formation [114]. Some of the molecules are documented that, it prevents the neurodegenerative process via alteration of the various cellular signaling process. Some of LPO peroxides are formed by lipoxygenase enzymatic system. Thus, 5-lipoxygenase (5-LOX) inhibitor, that is, zileuton employed as a reduction of LPO products in the biological system via ligation of the active site of iron through its N-hydroxy urea moiety [58]. In addition, the 12- and 15-lipoxygenases (12-LOX and 15-LOX) process contributes to the significant role in the pathogenesis of neurodegenerative disorders via accumulation of LPO products [115]. The discovery of 12-LOX and 15-LOX inhibitors is still developing stage. The isotope of deuterated of PUFA products are documented to prevents the LPO products via enzymatic and non-enzymatic manner [18, 115]. Elimination of peroxides: the limited quantity of LPO products formation and reduction is not harmful to the cellular system. Whereas, the abundant accumulation of LPO products makes the serious complication to the biological system via alternating cell signal and functions [116]. Generally, the biological system eliminates the LPO products via the enzymatic mechanism [117, 118]. The glutathione peroxidase (GPx) enzyme is one of the classical enzymes for reduction of LPO products. Total eight isoforms of GPx enzymes are distributed in humans with different substrate specificities and tissue specificity. Among all GPx enzymes, the GPx4 is identified as a primary enzyme for the reduction of lipid peroxides. In addition, it also interferes with ferrous ion leads to reduce the ferroptotic cell death and accumulation of toxic lipid peroxides [40]. Another cofactor, that is, selenocysteine also helps to enhance the GPx4 enzyme activity for reduction of lipid peroxides using due to its strong nucleophilic attack on the terminal oxygen of lipid peroxide [119]. This reaction helps to the reduction

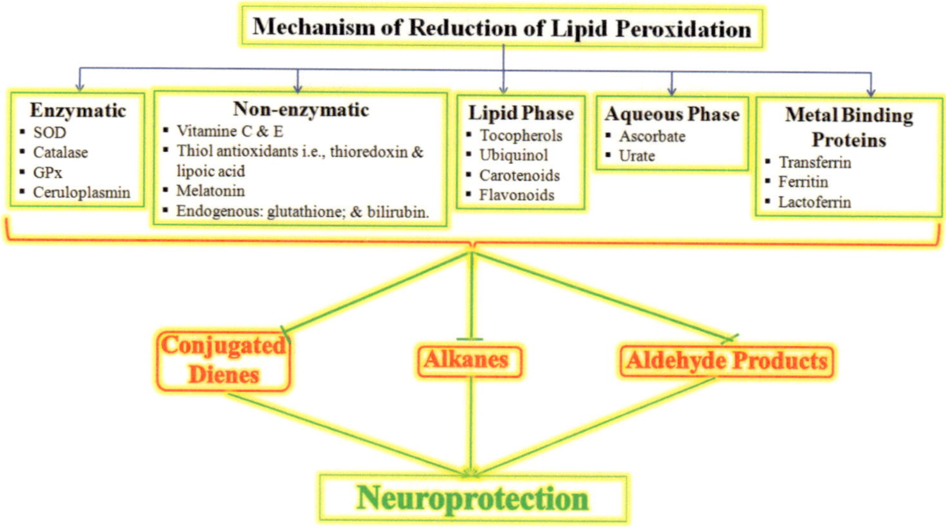

Figure 3.
Mechanism of reduction of lipid radicals. The primary event of lipid radicals is ROS formation. Thus, various antioxidant molecules are claimed as anti-lipid peroxidative agents. These agents are also known as free radical scavengers. This anti-oxidative molecules are categorized into the five ways, that is, enzymatic; non-enzymatic; lipid phase; solid phase actions; and activation of metal binding protein actions. The free radical scavenging molecules are reduce and termination the lipid radical in the biological system leads to neuroprotection from toxic radicals.

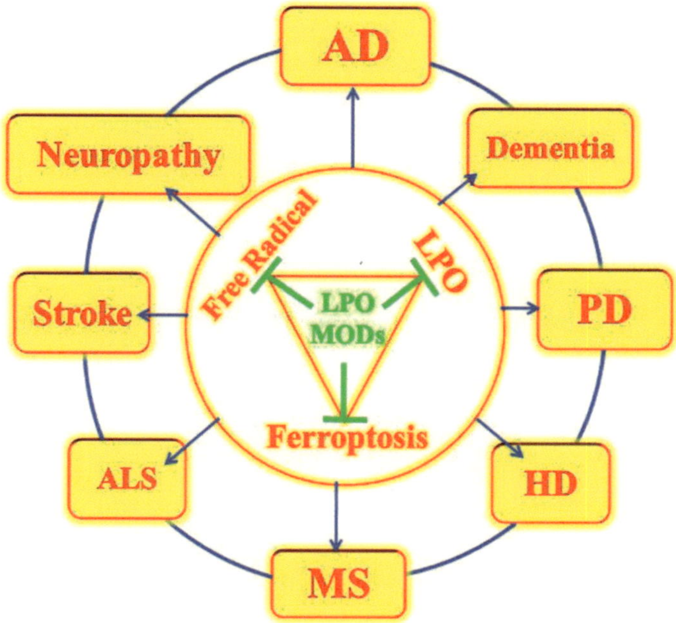

Figure 4.
Mechanism of LPO modulators for the prevention of neurodegenerative disorders. There are three major interrelated functions are inducing the neurodegeneration and neurological disorders such as (1) free radical generation; (2) LPO product accumulation; and (3) ferroptosis. The LPO modulators are attenuated the neurodegenerative disorders like anti-oxidant; anti-lipid peroxidation; and iron chelating actions.

of hydroxyeicosatetraenoic acid (HETE) or hydroxyoctadecadienoic acid (HODE) accumulation in the biological system. Furthermore, selenenic acid intermediates are reduced by the two molecules of oxidized glutathione and regenerate the active GPx enzyme [120]. List of LPO regulating molecules is listed in **Figures 3** and **4**.

15. Summary of LPO in neurodegenerative disorders

LPO products are contributes to the various types of neurodegenerative disorders such as AD, dementia, PD, HD, MS, ALS, stroke and neuropathy [108]. However, the availability of LPO products regulating molecules is limited. Their molecular action in *in-vivo* biological actions is not explored yet. Hence, the extensive researches are required to prove the potential ameliorative effect of LPO acting molecules to prevent the neurodegenerative disorders.

16. Future directions

Based on this complete literature and research reports, LPO regulating molecules has ample scope to prevent the neurodegenerative disorders. Because LPO products are documented to play a critical role in the various stage of the neurodegenerative disease. Even, it also interferes with multiple cell organelles such as mitochondria, endoplasmic reticulum, lysosome, peroxisome, nucleus and various cytoskeletal proteins. The numerous reports also documented that, the molecules are altering the LPO process and reducing the LPO products accumulations. Whereas, all the classical neuroprotective drugs are claimed as ion channel regulator; enzyme modulators; receptor antagonist actions so on. The direct and specific LPO pathway regulating molecules are not identified to attenuate the neurodegenerative disorders in vivo pharmacological research. Even, the newer concept of LPO associated ferroptosis actions enhances the neurodegeneration, but ferroptosis regulating molecules in the management of neurodegenerative disorders need to be study extensively. So, the discovery of LPO pathway modulating agents can treat the neurodegenerative disorders. Hence, we believe that, this book chapter will be helpful to the various researchers; who working on newer molecule discovery process for the prevention of LPO associated neurodegenerative disorders.

Acknowledgements

The author are thankful to Department of Pharmacology, JSS College of Pharmacy, JSS Academy of Higher Education and Research, Mysuru-570 015, Karnataka, India, for supporting and timely helps for the preparation of this book chapter with unconditional technical facilities.

Author details

Arunachalam Muthuraman[1*], Narahari Rishitha[1], Nallupillai Paramakrishnan[1], Bhaskaran Mahendran[1] and Muthusamy Ramesh[2]

1 JSS College of Pharmacy, JSS Academy of Higher Education and Research, Mysuru, Karnataka, India

2 Department of Pharmaceutical Analysis, Omega College of Pharmacy, Ghatkesar, Telangana, India

*Address all correspondence to: arunachalammu@gmail.com

References

[1] Nita M, Grzybowski A. The role of the reactive oxygen species and oxidative stress in the pathomechanism of the age-related ocular diseases and other pathologies of the anterior and posterior eye segments in adults. Oxidative Medicine and Cellular Longevity. 2016;**2016**:3164734

[2] Barone E et al. HNE-modified proteins in Down syndrome: Involvement in development of Alzheimer disease neuropathology. Free Radical Biology and Medicine. 2017;**111**:262-269

[3] Ayala A, Muñoz MF, Argüelles S. Lipid peroxidation: Production, metabolism, and signaling mechanisms of malondialdehyde and 4-hydroxy-2-nonenal. Oxidative Medicine and Cellular Longevity. 2014;**2014**:360438

[4] Sergeeva I. Structural-metabolic characteristics of cells and their functional opportunities. 2018:**1**(1):1-7

[5] Yang WS et al. Peroxidation of polyunsaturated fatty acids by lipoxygenases drives ferroptosis. Proceedings of the National Academy of Sciences. 2016;**113**(34):E4966-E4975

[6] Das K, Roychoudhury A. Reactive oxygen species (ROS) and response of antioxidants as ROS-scavengers during environmental stress in plants. Frontiers in Environmental Science. 2014;**2**:53

[7] Birben E et al. Oxidative stress and antioxidant defense. World Allergy Organization Journal. 2012;**5**(1):9

[8] Sheikh S, Haque E, Mir SS. Neurodegenerative diseases: Multifactorial conformational diseases and their therapeutic interventions. Journal of Neurodegenerative Diseases. 2012;**2013**:563481

[9] Perez-Matos M, Morales-Alvarez M, Mendivil C. Lipids: A suitable therapeutic target in diabetic neuropathy? Journal of Diabetes Research. 2017;**2017**:6943851

[10] Aprioku JS. Pharmacology of free radicals and the impact of reactive oxygen species on the testis. Journal of Reproduction & Infertility. 2013;**14**(4):158

[11] Nowak JZ. Oxidative stress, polyunsaturated fatty acids derived oxidation products and bisretinoids as potential inducers of CNS diseases: Focus on age-related macular degeneration. Pharmacological Reports. 2013;**65**(2):288-304

[12] Gutowski M, Kowalczyk S. A study of free radical chemistry: Their role and pathophysiological significance. Acta Biochimica Polonica. 2013;**60**(1):1-16

[13] Ortiz GG et al. Cellular and biochemical actions of melatonin which protect against free radicals: Role in neurodegenerative disorders. Current Neuropharmacology. 2008;**6**(3):203-214

[14] Phaniendra A, Jestadi DB, Periyasamy L. Free radicals: Properties, sources, targets, and their implication in various diseases. Indian Journal of Clinical Biochemistry. 2015;**30**(1):11-26

[15] Nicolson GL, Ash ME. Membrane lipid replacement for chronic illnesses, aging and cancer using oral glycerolphospholipid formulations with fructooligosaccharides to restore phospholipid function in cellular membranes, organelles, cells and tissues. Biochimica et Biophysica Acta (BBA)-Biomembranes. 2017;**1859**(9):1704-1724

[16] Rafieian-Kopaei M et al. Atherosclerosis: Process, indicators, risk factors and new hopes. International Journal of Preventive Medicine. 2014;**5**(8):927

[17] Gahalain N et al. Lipid peroxidation: An overview. International Journal of Pharmaceutical Sciences and Research. 2011;**2**(11):2757

[18] Raefsky SM et al. Deuterated polyunsaturated fatty acids reduce brain lipid peroxidation and hippocampal amyloid β-peptide levels, without discernable behavioral effects in an APP/PS1 mutant transgenic mouse model of Alzheimer's disease. Neurobiology of Aging. 2018;**66**:165-176

[19] Bhat AH et al. Oxidative stress, mitochondrial dysfunction and neurodegenerative diseases; a mechanistic insight. Biomedicine & Pharmacotherapy. 2015;**74**:101-110

[20] Di Domenico F, Tramutola A, Butterfield DA. Role of 4-hydroxy-2-nonenal (HNE) in the pathogenesis of alzheimer disease and other selected age-related neurodegenerative disorders. Free Radical Biology and Medicine. 2017;**111**:253-261

[21] Kurutas EB. The importance of antioxidants which play the role in cellular response against oxidative/nitrosative stress: Current state. Nutrition Journal. 2015;**15**(1):71

[22] Guengerich FP, Yoshimoto FK. Formation and cleavage of C—C bonds by enzymatic oxidation–reduction reactions. Chemical Reviews. 2018;**118**(14):6573-6655

[23] Maeda H et al. Tocopherols modulate extraplastidic polyunsaturated fatty acid metabolism in Arabidopsis at low temperature. The Plant Cell. 2008;**20**(2):452-470

[24] Yin H, Xu L, Porter NA. Free radical lipid peroxidation: Mechanisms and analysis. Chemical Reviews. 2011;**111**(10):5944-5972

[25] Bradley-Whitman MA, Lovell MA. Biomarkers of lipid peroxidation in Alzheimer disease (AD): An update. Archives of Toxicology. 2015;**89**(7):1035-1044

[26] Remello SN et al. Two-electron oxidation of water to form hydrogen peroxide catalysed by silicon-porphyrins. Sustainable Energy & Fuels. 2018;**2**(9):1966-1973

[27] Kagan VE. Lipid peroxidation. In: Biomembranes. New York: CRC Press; 2018

[28] Zoidis E et al. Selenium-dependent antioxidant enzymes: Actions and properties of selenoproteins. Antioxidants. 2018;**7**(5):66

[29] Matsunobu T et al. Thromboxane A synthase-independent production of 12-hydroxyheptadecatrienoic acid, a BLT2 ligand. Journal of Lipid Research. 2013;**54**(11):2979-2987

[30] Yang C, Shahidi F, Tsao R. Biomarkers of oxidative stress and cellular-based assays of indirect antioxidant measurement. Measurement of Antioxidant Activity and Capacity: Recent Trends and Applications. 2018. p. 165. ISBN: 978-1-119-13535-7

[31] Saieva C et al. Dietary and lifestyle determinants of malondialdehyde DNA adducts in a representative sample of the Florence City population. Mutagenesis. 2016;**31**(4):475-480

[32] Sathyasaikumar KV et al. Assessing and Modulating Kynurenine Pathway Dynamics in Huntington's Disease: Focus on Kynurenine 3-Monooxygenase, in Huntington's Disease. New York: Springer; 2018. pp. 397-413

[33] Farina M et al. Metals, oxidative stress and neurodegeneration: A focus on iron, manganese and mercury. Neurochemistry International. 2013;**62**(5):575-594

[34] Ijomone OM et al. Sub-acute nickel exposure impairs behavior, alters neuronal microarchitecture, and induces oxidative stress in rats' brain. Drug and Chemical Toxicology. 2018;**26**:1-8

[35] Kumar A et al. Biochemical and Molecular Targets of Heavy Metals and their Actions, in Biomedical Applications of Metals. Switzerland: Springer; 2018. pp. 297-319

[36] Lane DJ, Ayton S, Bush AI. Iron and alzheimer's disease: An update on emerging mechanisms. Journal of Alzheimer's Disease. 2018;**64**(S1):1-16

[37] Conrad M et al. Regulation of lipid peroxidation and ferroptosis in diverse species. Genes & Development. 2018;**32**(9-10):602-619

[38] Feng H, Stockwell BR. Unsolved mysteries: How does lipid peroxidation cause ferroptosis? PLoS Biology. 2018;**16**(5):e2006203

[39] Morris G et al. Why should neuroscientists worry about iron? The emerging role of ferroptosis in the pathophysiology of neuroprogressive diseases. Behavioural Brain Research. 2017;**341**:154-175

[40] Maiorino M, Conrad M, Ursini F. GPx4, lipid peroxidation, and cell death: Discoveries, rediscoveries, and open issues. Antioxidants & Redox Signaling. 2018;**29**(1):61-74

[41] Song Y et al. Polyunsaturated fatty acid relatively decreases cholesterol content in THP-1 macrophage-derived foam cell: Partly correlates with expression profile of CIDE and PAT members. Lipids in Health and Disease. 2013;**12**(1):111

[42] Sergent O, Morel I, Cillard J. Involvement of metal ions in lipid peroxidation: Biological implications. Metal Ions in Biological Systems. 1999;**36**:251-287

[43] Garza-Lombó C et al. Neurotoxicity linked to dysfunctional metal ion homeostasis and xenobiotic metal exposure: Redox signaling and oxidative stress. Antioxidants & Redox Signaling. 2018;**28**(18):1669-1703

[44] Yaribeygi H et al. The underlying role of oxidative stress in neurodegeneration: A mechanistic review. CNS & Neurological Disorders-Drug Targets. 2018;**17**(3):207-215

[45] Agrawal S et al. Brain mitochondrial iron accumulates in Huntington's disease, mediates mitochondrial dysfunction, and can be removed pharmacologically. Free Radical Biology and Medicine. 2018;**120**:317-329

[46] Cortes CJ, La Spada AR. TFEB dysregulation as a driver of autophagy dysfunction in neurodegenerative disease: Molecular mechanisms, cellular processes, and emerging therapeutic opportunities. Neurobiology of Disease. 2018;**18**:30150-30155

[47] Flanagan E et al. Impact of flavonoids on cellular and molecular mechanisms underlying age-related cognitive decline and neurodegeneration. Current Nutrition Reports. 2018;**7**(2):49-57

[48] Ansari MA, Scheff SW. Oxidative stress in the progression of Alzheimer disease in the frontal cortex. Journal of Neuropathology & Experimental Neurology. 2010;**69**(2):155-167

[49] Mecocci P et al. A long journey into aging, brain aging, and Alzheimer's disease following the oxidative stress tracks. Journal of Alzheimer's Disease. 2018;**62**(3):1319-1335

[50] Janocko NJ et al. Neuropathologically defined subtypes of Alzheimer's disease differ significantly from neurofibrillary tangle-predominant dementia. Acta Neuropathologica. 2012;**124**(5):681-692

[51] Choi ML, Gandhi S. Crucial role of protein oligomerization in the pathogenesis of Alzheimer's and Parkinson's diseases. The FEBS Journal. 2018. DOI: 10.1111/febs.14587. (Article in press)

[52] Gaj P et al. Identification of a late onset Alzheimer's disease candidate risk variant at 9q21.33 in Polish patients. Journal of Alzheimer's Disease. 2012;**32**(1):157-168

[53] Shinohara M et al. Regional distribution of synaptic markers and APP correlate with distinct clinicopathological features in sporadic and familial Alzheimer's disease. Brain. 2014;**137**(Pt 5):1533-1549

[54] Yuan Q et al. Association of polymorphisms in the LRP1 and A2M genes with Alzheimer's disease in the northern Chinese Han population. Journal of Clinical Neuroscience. 2013;**20**(2):253-256

[55] Vinothkumar G et al. Therapeutic impact of rHuEPO on abnormal platelet APP, BACE 1, presenilin 1, ADAM 10 and Aβ expressions in chronic kidney disease patients with cognitive dysfunction like Alzheimer's disease: A pilot study. Biomedicine & Pharmacotherapy. 2018;**104**:211-222

[56] Atwood CS et al. Role of free radicals and metal ions in the pathogenesis of Alzheimer's disease. Metal Ions in Biological Systems. 1999;**36**:309-364

[57] Kalász H et al. Pharmacognostical sources of popular medicine to treat Alzheimer's disease. The Open Medicinal Chemistry Journal. 2018;**12**:23

[58] Panigrahi S et al. 5-lipoxygenase: Emerging therapeutic targets in central nervous system disorders. International Journal of Advanced Research in Biological Sciences. 2018;**5**(3):20-29

[59] Verma SK et al. Enhancement in the neuroprotective power of riluzole against cerebral ischemia using a brain targeted drug delivery vehicle. ACS Applied Materials & Interfaces. 2016;**8**(30):19716-19723

[60] Butterfield DA, Lange MLB, Sultana R. Involvements of the lipid peroxidation product, HNE, in the pathogenesis and progression of Alzheimer's disease. Biochimica et Biophysica Acta (BBA)-Molecular and Cell Biology of Lipids. 2010;**1801**(8):924-929

[61] Than S, Srikanth V. Detecting brain injury related to hypertension at midlife—A key to interventions for preventing dementia in older age. Cardiovascular Research. 2018;**114**:1430-1431

[62] Kondo S, Kondo Y. Association of the types of dementia and the incidence of hypertension, diabetes mellitus, or bone fracture. Gan to kagaku ryoho. Cancer & Chemotherapy. 2018;**45**(Suppl 1):25-26

[63] Jordan BC et al. Dementia as a predictor of mortality in adult trauma patients. The American Journal of Surgery. 2018;**215**(1):48-52

[64] Zhang S et al. Intercellular transfer of pathogenic α-synuclein by extracellular vesicles is induced by the lipid peroxidation product 4-hydroxynonenal. Neurobiology of Aging. 2018;**61**:52-65

[65] Yadav A et al. Resveratrol loaded solid lipid nanoparticles attenuate mitochondrial oxidative stress in vascular dementia by activating Nrf2/HO-1 pathway. Neurochemistry International. 2018;**112**:239-254

[66] Bastías-Candia S, Zolezzi JM, Inestrosa NC. Revisiting the paraquat-induced sporadic Parkinson's disease-like model. Molecular Neurobiology. 2018. DOI: 10.1007/s12035-018-1148-z. (Article in press)

[67] Jaswal G et al. Reduced substantia innominata volume mediates contributions of microvascular and macrovascular disease to cognitive deficits in Alzheimer's disease. Neurobiology of Aging. 2018;**66**:23-31

[68] Ehara A et al. Role of neuronal nitric oxide synthase in slowly progressive dopaminergic neurodegeneration in the Zitter rat. Nitric Oxide. 2018;**78**:41-50

[69] Joe E-H et al. Astrocytes, microglia, and Parkinson's disease. Experimental Neurobiology. 2018;**27**(2):77-87

[70] Angelova PR et al. Lipid peroxidation is essential for α-synuclein-induced cell death. Journal of Neurochemistry. 2015;**133**(4):582-589

[71] Rachitha P et al. Chemical composition, antioxidant potential, macromolecule damage and neuroprotective activity of Convolvulus pluricaulis. Journal of Traditional and Complementary Medicine. 2018. DOI: ORG/10.1016/J.JTCME.2017.11.002. (Article in press)

[72] Lecumberri A et al. Neuronal density and proportion of interneurons in the associative, sensorimotor and limbic human striatum. Brain Structure and Function. 2018;**223**(4):1615-1625

[73] Esparza JL, Gómez M, Domingo JL. Role of melatonin in aluminum-related neurodegenerative disorders: A review. Biological Trace Element Research. 2018. DOI: 10.1007/s12011-018-1372-4. (Article in press)

[74] Vallée A et al. Aerobic glycolysis in amyotrophic lateral sclerosis and Huntington's disease. Reviews in the Neurosciences. 2018;**29**(5):547-555

[75] Jaglan D, Singh Bindra C. Evaluating neuroprotective effects of ascorbic acid against 3-nitropropionic acid induced huntington's disease in rats. Possible Involvement of Gabaareceptors. 2018;**6**(2):33-68

[76] Lee C-T et al. Current advances in the biological activity of polysaccharides in Dendrobium with intriguing therapeutic potential. Current Medicinal Chemistry. 2018;**25**(14):1663-1681

[77] Tully M et al. Systemic acrolein elevations in mice with experimental autoimmune encephalomyelitis and patients with multiple sclerosis. Frontiers in Neurology. 2018;**9**:420-429

[78] Sheykhansari S et al. Redox metals homeostasis in multiple sclerosis and amyotrophic lateral sclerosis: A review. Cell Death & Disease. 2018;**9**(3):348

[79] Elshaer ARM et al. Approach in diagnosis and management and common mistakes in diagnosis of multiple sclerosis. Egyptian Journal of Hospital Medicine. 2018;**70**(11):2008-2015

[80] Alfredsson L, Olsson T. Lifestyle and environmental factors in multiple sclerosis. Cold Spring Harbor Perspectives in Medicine. 2018. DOI: 10.1101/cshperspect.a028944. (Article in press)

[81] Campbell G, Mahad D. Neurodegeneration in progressive multiple sclerosis. Cold Spring Harbor Perspectives in Medicine. 2018. DOI: 10.1101/cshperspect.a028985. (Article in press)

[82] Pennetta G, Welte MA. Emerging links between lipid droplets and motor neuron diseases. Developmental Cell. 2018;**45**(4):427-432

[83] Hasegawa I et al. An autopsy case of globular glial tauopathy presenting with clinical features of motor neuron disease with dementia and iron deposition in the motor cortex. Neuropathology. 2018;**38**:372-379

[84] Komatsu K et al. Overexpressed wild-type superoxide dismutase 1

exhibits amyotrophic lateral sclerosis-related misfolded conformation in induced pluripotent stem cell-derived spinal motor neurons. Neuroreport. 2018;**29**(1):25-29

[85] Debette S et al. Association of MRI markers of vascular brain injury with incident stroke, mild cognitive impairment, dementia, and mortality: The Framingham Offspring Study. Stroke. 2010;**41**(4):600-606

[86] Krueger M et al. Blood-brain barrier breakdown after embolic stroke in rats occurs without ultrastructural evidence for disrupting tight junctions. PLoS One. 2013;**8**(2):e56419

[87] Gaignard P et al. Sex differences in brain mitochondrial metabolism: Influence of endogenous steroids and stroke. Journal of Neuroendocrinology. 2018;**30**(2):e12497

[88] Stecker M et al. Risk factors for DVT/PE in patients with stroke and intracranial hemorrhage. Open Neurology Journal. 2014;**8**:1-6

[89] Price AJ et al. Differences in risk factors for 3 types of stroke: UK prospective study and meta-analyses. Neurology. 2018;**90**(4):e298-e306

[90] Kong LL et al. Neutralization of chemokine-like factor 1, a novel C-C chemokine, protects against focal cerebral ischemia by inhibiting neutrophil infiltration via MAPK pathways in rats. Journal of Neuroinflammation. 2014;**11**:112

[91] Foreman PM et al. Endothelin polymorphisms as a risk factor for cerebral aneurysm rebleeding following aneurysmal subarachnoid hemorrhage. Clinical Neurology and Neurosurgery. 2017;**157**:65-69

[92] Vidale S et al. Post-ischemic inflammation in acute stroke. Journal of Clinical Neurology. 2017;**13**(1):1-9

[93] Goldstein LB et al. Primary prevention of ischemic stroke: A guideline from the American heart association/American stroke association stroke council: Cosponsored by the atherosclerotic peripheral vascular disease interdisciplinary working group; cardiovascular nursing council; clinical cardiology council; nutrition, physical activity, and metabolism council; and the quality of care and outcomes research interdisciplinary working group: The American academy of neurology affirms the value of this guideline. Stroke. 2006;**37**(6):1583-1633

[94] Tsikas D. Assessment of lipid peroxidation by measuring malondialdehyde (MDA) and relatives in biological samples: Analytical and biological challenges. Analytical Biochemistry. 2017;**524**:13-30

[95] Sun M-S et al. Free radical damage in ischemia-reperfusion injury: An obstacle in acute ischemic stroke after revascularization therapy. Oxidative Medicine and Cellular Longevity. 2018;**2018**:3804979

[96] Sayre LM, Perry G, Smith MA. Oxidative stress and neurotoxicity. Chemical Research in Toxicology. 2007;**21**(1):172-188

[97] Mrakic-Sposta S et al. R (+)-Thioctic acid effects on oxidative stress and peripheral neuropathy in type II diabetic patients: Preliminary results by electron paramagnetic resonance and electroneurography. Oxidative Medicine and Cellular Longevity. 2018;**2018**:1767265

[98] Hlusicka J et al. Role of activation of lipid peroxidation in the mechanisms of acute methanol poisoning. Clinical Toxicology. 2018. p. 1-11. DOI: 10.1080/15563650.2018.1455980

[99] Inoue K, Tsuda M. Microglia in neuropathic pain: Cellular and molecular mechanisms and

therapeutic potential. Nature Reviews Neuroscience. 2018;**19**(3):138-152

[100] Morris G et al. Cell death pathways: A novel therapeutic approach for neuroscientists. Molecular Neurobiology. 2018;**55**(7):5767-5786

[101] Rice AS et al. Sensory profiling in animal models of neuropathic pain: A call for back-translation. Pain. 2018;**159**(5):819-824

[102] Dewanjee S et al. Molecular mechanism of diabetic neuropathy and its pharmacotherapeutic targets. European Journal of Pharmacology. 2018;**833**:472-523

[103] Sengupta A et al. Balance between antioxidant enzyme and lipid peroxidation in diabetes neuropathy. Balance. 2018;**3**(2):1125-1128

[104] Apostolopoulou K et al. Ischemia–reperfusion injury of sciatic nerve in rats: Protective role of combination of vitamin C with E and tissue plasminogen activator. Neurochemical Research. 2018;**43**(3):650-658

[105] Muthuraman A, Ramesh M. Ischemic-reperfusion of unilateral external iliac artery in rat: A new model for vasculitic femoral neuropathy. Neuroscience Letters. 2016;**628**:10-16

[106] Muthuraman A, Singla SK, Peters A. Exploring the potential of flunarizine for cisplatin-induced painful uremic neuropathy in rats. International Neurourology Journal. 2011;**15**(3):127-134

[107] Frank B, Gupta S. A review of antioxidants and Alzheimer's disease. Annals of Clinical Psychiatry. 2005;**17**(4):269-286

[108] Sottero B et al. Lipid oxidation derived aldehydes and oxysterols between health and disease. European Journal of Lipid Science and Technology. 2018. p. 1700047. DOI: ORG/10.1002/EJLT.201700047

[109] Mirończuk-Chodakowska I, Witkowska AM, Zujko ME. Endogenous non-enzymatic antioxidants in the human body. Advances in Medical Sciences. 2018;**63**(1):68-78

[110] Mehta V et al. ACE Alzheimer's: The role of vitamin A, C and E (ACE) in oxidative stress induced Alzheimer's disease. Journal of Medical Research and Innovation. 2018;**2**(1):e000086-e000086

[111] Xiao M et al. Pathophysiology of mitochondrial lipid oxidation: Role of 4-hydroxynonenal (4-HNE) and other bioactive lipids in mitochondria. Free Radical Biology and Medicine. 2017;**111**:316-327

[112] Elharram A et al. Deuterium-reinforced polyunsaturated fatty acids improve cognition in a mouse model of sporadic Alzheimer's disease. The FEBS Journal. 2017;**284**(23):4083-4095

[113] Zarkovic K, Jakovcevic A, Zarkovic N. Contribution of the HNE-immunohistochemistry to modern pathological concepts of major human diseases. Free Radical Biology and Medicine. 2017;**111**:110-126

[114] Mashima R, Maekawa M. Lipid biomarkers for the peroxisomal and lysosomal disorders: Their formation, metabolism and measurement. Biomarkers in Medicine. 2018;**12**(1):83-95

[115] Li Q-Q et al. 12/15 lipoxygenase: A crucial enzyme in diverse types of cell death. Neurochemistry International. 2018;**118**:34-41

[116] Sbodio JI, Snyder SH, Paul BD. Redox mechanisms in neurodegeneration: From disease

outcomes to therapeutic opportunities. Antioxidants & Redox Signaling. 2018. DOI: ORG/10.1089/ARS.2017.7321. (Article in press)

[117] Truong VL, Jun M, Jeong WS. Role of resveratrol in regulation of cellular defense systems against oxidative stress. BioFactors. 2018;**44**(1):36-49

[118] Maher P. Potentiation of glutathione loss and nerve cell death by the transition metals iron and copper: Implications for age-related neurodegenerative diseases. Free Radical Biology and Medicine. 2018;**115**:92-104

[119] Borchert A et al. Crystal structure and functional characterization of selenocysteine-containing glutathione peroxidase 4 suggests an alternative mechanism of peroxide reduction. Biochimica et Biophysica Acta (BBA)-Molecular and Cell Biology of Lipids. 2018;**1863**(9):1095-1107

[120] Lenardão EJ, Santi C, Sancineto L. Bioactive organo selenium compounds and therapeutic perspectives. In: New Frontiers in Organoselenium Compounds. Switzerland: Springer; 2018. ISBN(P): 978-3-319-92404-5

Chapter 3

Lipid Peroxidation in Meat and Meat Products

Ana Lúcia F. Pereira and Virgínia Kelly G. Abreu

Abstract

The meat and meat products present a considerable amount of lipid in their composition. The lipid composition of these foods is diversified. Thus, depending on the type of meat, which can be rich in unsaturated fatty acid, there is an increase in the disposition for lipid oxidation. Oxidation reactions not only reduce the shelf life and nutritional value of food products but also can generate harmful compounds. Thus, having in view that many types of new technologies are applied to these foods, the proposal of this chapter of how these new methodologies have affected the lipid peroxidation of these foods. Moreover, the aim is to evaluate what impacts on the chemical characteristics of these foods.

Keywords: fatty acids, rancidity, meat preservation, high pressure, microwave heating, ultraviolet light, infrared heating, radiation

1. Introduction

Oxidation is one of the essential factors in the nonmicrobial degradation of meat and meat products. Thus, the lipid oxidation has been extensively investigated in these foods because the products of the reaction can readily react with proteins, leading to sensory modifications and the loss of nutritional value [1].

Food preservation is a process to extend the shelf life of foods while maintaining their safety and sensory properties. Nowadays, some new preservation techniques are being developed to satisfy the current demands for more efficient preservation and higher consumer satisfaction about nutritional and sensory aspects, convenience, safety, absence of chemical preservatives, low price, and environmental safety [2]. These methods include high-pressure processing, microwave heating, ultraviolet light, infrared heating, and radiation.

The high-pressure processing (HPP) is considered a food safety process that can stabilize meat by inactivating microorganisms. However, HPP can favor the lipid peroxidation by promoting the formation of radicals [1].

In the microwave heating processing, changes associated with chemical components of food products relate mainly to the cooking loss, antioxidant activity, bioactive components, and lipid peroxidation. During microwave cooking, protein denaturation, cooking loss, and lipid peroxidation of meat and meat products increase with the increase in heating time or temperature [3].

Another technique entirely used in research is the infrared cooking that consists of the penetration of electromagnetic waves to the food material. The absorbed infrared waves could cause electromagnetic vibrations and result in temperature increase within the food material. The penetration capacity of infrared waves limits the whole cooking of food material [4]. This technology is of particular interest to the processed meat sector, since conventional cooking ovens using high-velocity hot air convection can cause overheating, oxidation, charring, impingement damage, low yield, difficult emissions, as well as high energy costs. Infrared radiation has intrinsic advantages such as having no direct intention or necessity to heat the air, keeping oven temperatures and humidity at low values. A further advantage of this method is the ease with which heat can be applied evenly over a broad surface area [5].

Radiation from ultraviolet light C (UV-C) also has been demonstrated as a potential surface decontamination method in addition to several advantages over regular sanitation methods. However, UV-C radiation possibly affects the physicochemical properties of meat products [6]. Paskeviciute et al. [7] used the UV-C for decontamination of chicken from food pathogens and observed small changes in the intensity of lipid peroxidation (0.16 mg malondialdehyde per kilogram of chicken meat). Moreover, these authors reported that sensory properties of treated chicken did not have changes of raw chicken, chicken broth, or cooked chicken meat when treated under nonthermal conditions in comparison with control.

Finally, the food radiation is one of the nonthermal methods of meat preservation. It is the process of exposing the food, either in the package or in bulk, to controlled amounts of ionizing radiation to achieve a purpose such as the extension of shelf-life, insect disinfection, the elimination of food-borne pathogens, and parasites [8]. It is considered a more effective and appropriate method to enhance food stability and safety when compared to other processing methods like heat and chemical methods. Also, it does not reduce significantly the nutritional and the sensory quality of food at lower doses. According to Fallah et al. [9], gamma irradiation had no significant effect on the primary sensory attributes of the irradiated samples of ready-to-cook Iranian barbecued chicken. Moreover, at the end of the storage period of 15 days, the irradiated samples had more sensory acceptance than nonirradiated samples.

Therefore, considering that most of the new technologies mentioned increase lipid peroxidation, strategies have been adopted to reduce this process. Among the measures adopted, include vacuum packaging and use of antioxidants [10].

2. Lipid composition in meat and meat products

Meat lipids are mainly composed of triglycerides (correspond to about 95% of meat lipids) and phospholipids, which contain saturated fatty acids, monounsaturated fatty acids (MUFAs), and polyunsaturated fatty acids (PUFAs). Triglycerides are storage lipids and are composed of three fatty acids esterified to glycerol and have more ratios of saturated fatty acids. The phospholipids are often functional lipids prevalent in cell membranes and as such contain more PUFA than triglycerides [11]. The main unsaturated fatty acids present in meat lipids have one or two double-chain linkages. The most common is the monounsaturated oleic acid (C18:1), which corresponds to about 40% of the fatty acids in beef. The two main PUFAs, linoleic (LA, C18:2 n − 6) and linolenic (C18:3 n − 3), form a substantial part of the membrane lipids but also form part of the storage lipids [12].

The firmness fat meat depends on the amount of saturated fatty acids. In general, the cattle fat is more saturated than those of pigs, and these are more saturated than those of poultry. Thus, the saturated fatty acids content explains the higher hardness of fat in this sequence, cattle > pigs > poultry. The melting point of cattle fat is between 43 and 47°C, while that of pigs is between 38 and 44°C and that of poultry is between 31 and 37°C [13].

According to Wood et al. [14], the fatty acid composition of adipose tissue and muscle in pigs, sheep, and cattle depends on the amount of fat in the carcass and

	Adipose tissue			Muscle		
	Pigs	Sheep	Cattle	Pigs	Sheep	Cattle
14:0	1.6[a]	4.1[b]	3.7[b]	1.3[a]	3.3[c]	2.7[b]
16:0	23.9[b]	21.9[a]	26.1[c]	23.2[b]	22.2[a]	25.0[c]
16:1 cis	2.4[a]	2.4[a]	6.2[b]	2.7[b]	2.2[a]	4.5[c]
18:0	12.8[a]	22.6[b]	12.2[a]	12.2[a]	18.1[c]	13.4[b]
18:1 cis – 9	35.8[b]	28.7[a]	35.3[b]	32.8[a]	32.5[a]	36.1[b]
18:2 n – 6	14.3[b]	1.3[a]	1.1[a]	14.2[b]	2.7[a]	2.4[a]
18:3 n – 3	1.4[c]	1.0[b]	0.5[a]	0.95[b]	1.37[c]	0.70[a]
20:4 n – 6	0.2	ND	ND	2.21[b]	0.64[a]	0.63[a]
20:5 n – 3	ND	ND	ND	0.31[b]	0.45[c]	0.28[a]
n – 6:n – 3	7.6	1.4	2.3	7.2	1.3	2.1
P:S	0.61	0.09	0.05	0.58	0.15	0.11
Total	65.3	70.6	70.0	2.2	4.9	3.8

Means with different superscripts (a, b, and c) are significantly different ($p < 0.05$).

Table 1.
Fatty acid composition (g/100 g fatty acids) and content (g/100 g total fatty acids in subcutaneous adipose tissue and muscle) of loin steaks/chops in pigs, sheep, and cattle [15].

muscle. **Table 1** shows that adipose tissue has a much higher fatty acid content than muscle, but the fatty acid composition of the two tissues is broadly similar. However, there are significant species differences. Pigs have much higher proportions of the polyunsaturated fatty acid (PUFA) linoleic acid (18:2 n – 6) in both tissues than cattle and sheep.

The linoleic acid (18:2 n – 6) derives from the diet. In pig, it passes through the stomach to have changed and is then absorbed and incorporated from there into tissues. In ruminants, the fatty acids which are at high levels in concentrate feedstuffs are degraded into MUFAs in the rumen by biohydrogenation and only a small proportion, around 10% of dietary 18:2 n – 6, is available for incorporation into tissue lipids. In the sheep and cattle, the fatty acid is at higher levels in muscle than adipose tissue. The second most important PUFA is α-linolenic acid (18:3 n – 3), which is present in many concentrate feed ingredients but at lower levels than 18:2 n – 6. In pigs, the proportion is higher in adipose tissue than muscle. The linolenic acid is a primary dietary fatty acid for ruminants, since it constitutes over 50% of total fatty acids in the grass [13, 16, 17].

Greater incorporation of 18:2 n – 6 into pig muscle fatty acids compared with ruminants produces higher levels of 20:4 n – 6 by synthesis, and the net result is a higher ratio of n – 6:n – 3 PUFA compared with the ruminants (**Table 1**). Nutritional advice is for ratios <4.0 [18], and so pig muscle is unbalanced relative to that of the ruminants. On the other hand, the ratio of all PUFAs to saturated fatty acids (P:S), the target for which is 0.4 or above, is much higher, beneficially so, in pigs and other monogastrics compared with the ruminants [14].

Moreover, in beef, conjugated linoleic acids (CLAs) are produced in the rumen by biohydrogenation at a level of approximately 1.2e10 mg per g of fat, which results in approximately 36 mg of linoleic acid per g of fat [19, 20]. Studies reported that minced meat (15% lipids) had an average content of 120 mg CLA per 100 g of steak. Moreover, the animal's diet can influence the CLA content of beef meat [1].

3. Lipid peroxidation in meat and meat products

The lipid peroxidation is a primary reason for the deterioration of meat and meat products, giving undesirable odors, rancidity, texture modification, loss of essential fatty acids, or toxic compound production. Moreover, lipid oxidation products implicate several human pathologies (atherosclerosis, cancer, inflammation, or aging processes) [21, 22].

Lipid peroxidation generally involves the degradation of polyunsaturated fatty acids (PUFAs) and the production of secondary decomposition products, including carbonyls and hydrocarbon compounds. The oxidative stability of meat depends on the balance of anti- and pro-oxidants and the composition of oxidizable substrates, including PUFAs, cholesterol, proteins, and pigments [23–26].

The peroxidation reaction of PUFA in biological tissues can be initiated by free radicals, which are present in animal cells with active metabolic processes. After the slaughter animals, their muscle cells become overloaded with pro-oxidants, peroxidized lipids, and oxygen radicals. These changes also occur during storage at 2–4°C. The secondary phase of lipid peroxidation should occur immediately after slaughtering and occurs during the early postslaughter phase. The biochemical changes, which accompany the conversion of muscle to meat, generate conditions in which the oxidation in the highly unsaturated phospholipid fraction of subcellular membranes is no longer tightly controlled [27, 28].

Malondialdehyde (MDA), which is a three-carbon compound formed after the scission of peroxidized PUFAs, is one of the main products of lipid peroxidation. Consequently, there is the aldehydes production in substantial quantities during lipid oxidation, and therefore, these compounds are candidates for reactions with thiobarbituric acid (TBA). Consequently, the detection of these secondary products through chemical or instrument methods is relevant in studies examining lipid peroxidation in meat and meat products [29–31].

Many methods have been proposed for evaluating the MDA content in meat as a marker of lipid oxidation. High-performance liquid and gas chromatographic methods offer better specificity and sensitivity when detecting malondialdehyde. However, spectrophotometric methods are preferable during routine analyses of large samples due to their simplicity and low cost [32, 33]. The TBA test is the most common method used to quantify lipid oxidation products through the determination of MDA [34].

4. New technologies for preserving meat products and their impacts on lipid stability

More than two decades ago, novel food processing technologies that based on high tech or cutting-edge advances started to emerge to address productivity issues, extending product shelf life without affecting the nutritional content, sensory attributes, and product specifications. In research performed with food professionals from industry, academia and government observed that technologies such as high-pressure processing, microwave heating, ultraviolet light, infrared heating, and radiation were scored well for implementation or potential implementation in meat sector [35].

High-pressure processing (HPP) is also called high hydrostatic pressure processing, pascalization, or high-pressure pasteurization. This technology effectively inactivates vegetative bacteria, yeast, and molds using pressures up to 600 MPa at ambient temperature and can inactivate spores when combined with high temperature (high-pressure thermal processing). Moreover, the HPP retains most of the sensory and nutritional characteristics of solid or chilled products. Its effect on enzymes is variable [36–38].

HPP treatment, when used in meat, can promote peroxidation reactions, and it is essential to control the balance between pro-oxidants and antioxidants to prevent this phenomenon. Thus, many researchers have been interested in evaluating the extent of oxidation in pressurized meat to understand the underlying mechanisms. In particular, the fate of proteins such as myoglobin and hemoglobin under high-pressure treatment has been investigated because these proteins act as pro-oxidants in raw meat [1].

Studies concluded that treatment at pressures above 350 MPa has a pro-oxidant effect for all types of meat [38–44]. Moreover, the lipid peroxidation levels have been evaluated during storage after HPP treatment. After treatment at pressures between 300 and 800 MPa, the TBARS values increased in chicken meat kept at 5°C for 14 days, mainly when used at more than 400 Mpa [45]. Similar results were reported for beef pressurized at between 200 and 600 MPa for 20 min and kept refrigerated for 7 days [41].

The lipid peroxidation in meat products, such as dried products, is different than it is for raw or cooked meat due to the postprocessing operations and the longer conservation time. If HPP provides a prooxidant impact on meat products, this effect can be stressed by the subsequent storage [46]. Thus, the difference in the oxidative stability of dry-cured Iberian ham after 39 days of refrigerated storage for slices that had been pressurized at 400 MPa was observed. Treatment at 400 MPa led to discoloration of the products [47]. Moreover, HPP treatment of dry-cured loin after ripening affected its quality. However, using vacuum storage minimized the differences [39].

Thus, the extent of lipid peroxidation depends on the treatment duration, the temperature of the HPP treatment, and mainly on the type of meat or meat product. The beef seems to oxidize less than other types of meat. Furthermore, the initial packaging of the treated sample has a significant impact on the meat's peroxidation during the HPP treatment. Indeed, vacuum packaging, which is most frequently used for HPP treatment, reduces the impact of the pressure on the peroxidation process [42, 48, 49].

The microwave heating refers to the use of electromagnetic energy at the particular frequencies of 915 and 2450 MHz to generate heat in food. Contrary to conventional thermal techniques, heat is generated volumetrically throughout the product at faster rates. It can be used on solid foods [36, 37].

According to Byrnea et al. [50], for the control oxidation of any given lipid, the most critical parameters are the thermal treatment conditions (temperature and time of cooking). The lower temperature of cooking could reduce energy consumption, but a final internal temperature of 65–85°C must be reached to ensure safety [51]. Das and Rajkumar [52] evaluated the effects of various fat levels (5, 10, 15 and 20%) on microwave cooked goat meat patties. Each patty was cooked by microwave (700 W, 2.45 GHz) to an internal temperature of 75–80°C. Microwave cooking time was found to decrease with an increase in fat level, as the dielectric constant and loss factor decrease with fat content. Also, a sample with high-fat content might possess a lower specific heat capacity, which might lead to a decrease in the heating rate. The product yield (i.e., ratio of cooked weight to the raw weight) was found to be significantly lower for 20% fat level due to high total cooking loss (15.2%). Thus, the amount of fat content in food materials influences the microwave heating regarding heating rate, uniformity of temperature distribution, and fat retention [53].

Serrano et al. [54] reported that cooking methods, such as microwaves and conventional oven, did not increase TBARs values in restructured meat products. However, Dominguez et al. [55], comparing different cooking methods (roasting, grilling, microwaving, and frying) in the foal meat, observed that all the cooking methods increased TBARs content since high temperature during cooking causes increased oxidation in foal steaks. This increase was higher when foal steaks were microwaved or roasted. Therefore, many factors are influencing the lipid peroxidation when used in this technology.

The ultraviolet light-C (UV-C) produces nonionizing radiation with germicidal properties at wavelengths in the range of 200–280 nm. It can be used for surface treatment and as a nonthermal alternative [56, 57]. Different from thermal processing, this nonthermal technology reduces the microbial load without significantly changing the nutritional and sensory characteristics of meat products [58].

The beneficial effect of UV-C light on chicken meat was evaluated by many authors [59, 60]. These authors reported that UV light efficiently decreased the pathogenic bacterial load on the carcass surface without negatively affecting carcass color or meat lipid oxidation. On the other hand, Koutchma et al. [61] observed that UV light potentially affects food products due to free radical generation via a wide variety of organic photochemical reactions. Possible undesirable effects include oxidation of vitamins, lipids and proteins, degradation of antioxidants, changes in texture and color, and formation of off-flavors and aromas.

Lazaro et al. [6] evaluated three levels of UV-C intensities (0.62, 1.13, and 1.95 mW/cm^2) for up to 120 s in chicken breast. These authors reported that the intensity of 1.95 mW/cm^2 decreased the levels of pathogenic bacteria and can be used as a nonthermal technology to improve the superficial quality of packed poultry meat without promoting relevant changes on some quality indicators.

Infrared heating (IR) refers to the heating of materials by electromagnetic radiation having a wavelength of 1.3–4.0 μm (infrared radiation). This technique is based on the ability of materials to absorb a specific part of the spectrum of such radiation. Deep or superficial heating of the irradiated body, as well as local drying without heating the entire object, may be accomplished with appropriate selection of the emission spectrum of infrared radiation [62].

Shorter wavelength radiation penetrates the surface of meat and meat products more efficiently probably because of the preferential absorption spectrum of water in the surface. However, as the surface dries, this mechanism soon becomes less effective. Because of the higher fat content of the surface of the meat product samples, a higher proportion of the available infrared radiation from a far infrared source will be absorbed [5]. For the characteristics of the meat and meat products, the rapid action of surface heating provided by this method retains internally, flavor, aroma, and moisture, occurring changes only in the surface components, which may favor the Maillard reaction [63].

Turp et al. [64] evaluated the influence of final infrared cooking on characteristics of ohmically precooked meatballs. These authors concluded that infrared cooking, which is mainly useful for surface heating, can be applied as a final cooking method to improve the quality characteristics of ohmically precooked beef meatballs. Moreover, according to the same authors, the intensity of the infrared energy is affected by both the power applied and the distance between the infrared source and the meatball surface. Since the application of the different infrared intensities changes the total heat generation on the meatball surface, the temperature increase can vary.

The radiation technology includes irradiation by any of the three sources: gamma-rays, X-rays, or electron beams. They are often also referred to as ionizing radiations. Gamma rays can penetrate the food, but electron beams have a limited penetration depth [65, 66].

The radiation provides the production of free radicals, which could cause the changes in food components, such as the lipids and proteins in meat. Therefore, although radiation is a very useful cold sterilization technique, its utilization in meat or meat products has provided these problems. Studies have demonstrated that many chemical changes and quality changes in radiated meat were associated with free radical reactions, such as lipid and protein oxidation, which consequently caused the odor and color changes of meat [67]. Jo and Ahn [68] suggested that the lipolysis and lipid oxidation by the radiation played the critical role in the off-odor formation of irradiated meat.

Irradiation-induced oxidative chemical changes are dose dependent, and the presence of oxygen has a significant effect on the rate of oxidation of lipids and myoglobin in the muscle system [69]. Kim et al. [70] reported that no significant changes in the thiobarbituric acid reactive substances values (TBARS) of dry fermented sausages irradiated at 2 and 4 kGy during refrigerated storage. However, Kang et al. [71] showed that the levels of TBARS in irradiated half-dried seafood products increased as the dose was increased (from 3 to 10 kGy).

5. Strategies to reduce lipid peroxidation

For reduction of lipid peroxidation in meat and meat products, the antioxidant compounds have been added to products derived directly or, in some cases, incorporated into the diet of the animals. In recent years, special attention has been paid to some medicinal plants that could be used as potential sources of antioxidants for meat and meat products preservation and nutritional quality improvement. Most of the plant materials (herbs and spices) possess relatively high chemical nutrients (such as protein, fat, and carbohydrate), mineral contents (calcium, potassium, iron, phosphorus), and less anti-nutritional properties [72].

Pindi et al. [73] reported that *Kappaphycus alvarezii* (edible seaweed rich in polyphenolic substances) added in sausages reduced the lipid oxidation this poultry product during storage for 12 days at the 4°C. Panda and Cherian [74] reported that extracts of *Artemisia annua* (20 g. kg^{-1}) added in the broiler diets were useful in the lipidic oxidation delay of the poultry meat.

Bolumar et al. [75] evaluated the effect of the use of active antioxidant packaging for chicken meat processed by high-pressure treatment. For this, patties made of minced chicken breast and thigh packed in standard vacuum-packaging or active antioxidant packaging were subjected to high-pressure treatment (800 MPa, 10 min, 5°C) and subsequently stored for 25 days at 5°C. Lipid oxidation was studied at the surface and the inner parts of the meat patties. These authors observed that lipid oxidation was higher in the surface part, and the active packaging was able to delay it up to 25 days. The lipid oxidation was limited in the inner part of the meat patties and restrained at the surface of the active packaging.

According to Brewer [10], the methods to decrease the detrimental effects of irradiation include oxygen exclusion (vacuum packaging), replacement with inert gases (nitrogen), and the addition of protective agents (antioxidants).

Thus, Badr and Mahmoud [76] assessed the antioxidant effect of carrot juice in gamma-irradiated beef sausage. These authors observed that carrot juice significantly decreased the oxidative processes in the samples proportionally to the juice's concentration. Furthermore, the sausages that were formulated with carrot juice had a high acceptable sensory score as compared with the control samples.

6. Conclusions

Most of the methodologies used for the preservation of meats and meat products provide an increase in lipid peroxidation. The lipid peroxidation is one of the major factors limiting the quality and acceptability of meats and meat products. Among the technologies more promising to be used in meat and meat products has highlight the high-pressure processing, microwave heating, ultraviolet-light, infrared heating, and ionizing radiation.

For the high-pressure processing, studies have shown that this technology has a negative impact on lipid oxidation. This impact is generally limited by the antioxidants in the meat and by vacuum packaging.

Microwave heating of meat and meat products needs to be carried out to a great extent at a pilot scale level than at laboratory conditions so that the results might be useful for industrial applications. Despite the complex nature of microwave-food interactions, more research needs to be carried out for a better understanding of the process. Microwave heating products have the advantages of retaining more taste, color, quality, and nutritional value compared to those cooked by other conventional methods. This process is affected by the presence of moisture and fat content in food. Thus, in the literature, there are results which found increase or reduction in the lipid peroxidation depending on meat and meat product type, among others factors.

The ultraviolet-light is one such nonthermal technology that is approved for surface treatment of food, being an alternative surface decontaminant to be used for inactivating bacteria and viruses. Though with some limitations, if complemented with other processing techniques, this technology can help in better food preservation with minimal effects on the food quality.

Infrared heating offers many advantages over convection heating, including higher energy efficiency, heat transfer rate, and heat flux that results in time-saving as well as increased production line speed. This technology is attractive primarily for surface heating applications. In order to achieve energy optimum and efficient practical applicability of IR heating in the food processing industry, a combination of IR heating with microwave and other common conductive and convective modes of heating holds great potential.

The effects of ionizing radiation on meat could be reduced by various combinations of preslaughter feeding of antioxidants to livestock, the condition of the meat before irradiation (pH, oxymyoglobin vs. metmyoglobin), the addition of antioxidants directly to the product, gas atmosphere vacuum, packaging and temperature control.

Conflict of interest

There is no conflict of interest.

Author details

Ana Lúcia F. Pereira* and Virgínia Kelly G. Abreu
Federal University of Maranhão, Imperatriz, MA, Brazil

*Address all correspondence to: anafernandesp@gmail.com

References

[1] Guyon C, Meynier A, Lamballerie M. Protein and lipid oxidation in meat: A review with emphasis on high pressure treatments. Trends in Food Science and Technology. 2016;**50**:131-143

[2] Rahman MS. Food preservation and processing methods. In: Ahmed J, Rahman MS, editors. Handbook of Food Process Design. New York: Blackwell Publishing; 2012. pp. 1-17

[3] Guo Q, Sun D-W, Cheng J-H, Han Z. Microwave processing techniques and their recent applications in the food industry. Trends in Food Science and Technology. 2017;**67**:236-247

[4] Kor G, Icier F. Thermal imaging during infrared final cooking of semi-processed cylindrical meat product. Infrared Physics & Technology. 2016;**79**:242-251

[5] Sheridan P, Shilton N. Application of far infra-red radiation to cooking of meat products. Journal of Food Engineering. 1999;**41**:203-208

[6] Lazaro CA, Conte-Junior CA, Monteiro MLG, Canto ACVS, Costa-Lima BRC, Mano SB, et al. Effects of ultraviolet light on biogenic amines and other quality indicators of chicken meat during refrigerated storage. Poultry Science. 2014;**93**:2304-2313

[7] Paskeviciute E, Buchovec I, Luksiene Z. High power pulsed light for decontamination of chicken from food pathogens: A study on antimicrobial efficiency and organoleptic properties. Journal of Food Safety. 2011;**31**:61-68

[8] Shah MA, Mir SA, Pala SA. Enhancing food safety and stability through irradiation: A review. Journal of Microbiology, Biotechnology and Food Sciences. 2014;**3**:371-378

[9] Fallah AA, Saei-Dehkordi SS, Rahnama M. Enhancement of microbial quality and inactivation of pathogenic bacteria by gamma irradiation of ready-to-cook Iranian barbecued chicken. Radiation Physics and Chemistry. 2010;**79**:1073-1078

[10] Brewer MS. Irradiation effects on meat flavor: A review. Meat Science. 2009;**81**:1-14

[11] Mapiye C, Aldai N, Turner TD, Aalhus JL, Rolland DC, Kramer JKG, et al. Review—The labile lipid fraction of meat: From perceived disease and waste to health and opportunity. Meat Science. 2012;**92**:210-220

[12] Hwang Y-H, Joo S-T. Fatty acid profiles, meat quality, and sensory palatability of grain-fed and grass-fed beef from Hanwoo, American, and Australian crossbred cattle. Korean Journal for Food Science of Animal Resourches. 2017;**37**:153-161

[13] Gomide LAM, Ramos EM, Fontes PR. Ciência e Qualidade da Carne. 1st ed. Viçosa: UFV; 2013. 197p

[14] Wood JD, Enser M, Fisher AV, Nute GR, Sheard PR, Richardson RI, et al. Fat deposition, fatty acid composition and meat quality: A review. Meat Science. 2008;**78**:343-358

[15] Enser M, Hallett K, Hewitt B, Fursey GAJ, Wood JD. Fatty acid content and composition of English beef, lamb and pork at retail. Meat Science. 1996;**42**:443-456

[16] Ahmed ST, Islam M, Bostami ABMR, Mun H-S, Kim Y-J, Yang C-J. Meat composition, fatty acid profile and oxidative stability of meat from broilers supplemented with pomegranate (*Punica granatum* L.) by-products. Food Chemistry. 2015;**188**:481-488

[17] Oliveira EA, Sampaio AAM, Henrique W, Pivaro TM, Rosa BL, Fernandes ARM, et al. Quality traits and lipid composition of meat from Nellore young bulls fed with different oils either protected or unprotected from rumen degradation. Meat Science. 2012;**90**:28-35

[18] Scollan ND, Hocquette J-F, Nuernberg K, Dannenberger D, Richardson RI, Maloney A. Innovations in beef production systems that enhance the nutritional and health value of beef lipids and their relationship with meat quality. Meat Science. 2006;**74**:17-33

[19] Aharoni Y, Orlov A, Brosh A. Effects of high-forage content and oilseed supplementation of fattening diets on conjugated linoleic acid (CLA) and trans fatty acids profiles of beef lipid fractions. Animal Feed Science and Technology. 2004;**117**:43-60

[20] Schmid A, Collomb M, Sieber R, Bee G. Conjugated linoleic acid in meat and meat products: A review. Meat Science. 2006;**73**:29-41

[21] Alfaia CPM, Alves SP, Lopes AF, Fernandes MFE, Costa ASH, Fontes CMGA, et al. Effect of cooking methods on fatty acids, conjugated isomers of linoleic acid and nutritional quality of beef intramuscular fat. Meat Science. 2010;**84**:769-777

[22] Broncano JM, Petrón MJ, Parra V, Timón ML. Effect of different cooking methods on lipid oxidation and formation of free cholesterol oxidation products (COPs) in Latissimus dorsi muscle of Iberian pigs. Meat Science. 2009;**83**:431-437

[23] Brettonnet A, Hewavitarana A, Dejong S, Lanari MC. Phenolic acids composition and antioxidant activity of canola extracts in cooked beef, chicken and pork. Food Chemistry. 2010;**121**:927-933

[24] Lorenzo JM, Pateiro M. Influence of fat content on physicochemical and oxidative stability of foal liver pate. Meat Science. 2013;**95**:330-335

[25] Martino G, Haouet MN, Marchetti S, Grotta L. Effect of vitamin E supplementation on egg yolk quality and oxidative stability. Asian Journal of Agriculture and Food Sciences. 2014a;**2**:248-255

[26] Martino G, Mugnai C, Compagnone D, Grotta L, Del Carlo M, Sarti F. Comparison of performance, meat lipids and oxidative status of pigs from commercial breed and organic crossbreed. Animals. 2014b;**4**:348-360

[27] Martino G, Grotta L, Ponzielli V. Influence of dehydrated *Medicago sativa* on quality characteristics of Marchigiana beef. Animal Review. 2014c;**1**:37-44

[28] Min B, Ahn DU. Mechanism of lipid peroxidation in meat and meat products. Food Science and Biotechnology. 2005;**14**:152-163

[29] Fernández J, Pérez-àlvarez JA, Fernàndez-Lòpez JA. Thiobarbituric acid test for monitoring lipid oxidation in meat. Food Chemistry. 1997;**59**:345-353

[30] Gómez M, Lorenzo JM. Effect of fat level on physicochemical, volatile compounds and sensory characteristics of dry-ripened Bchorizo from Celta pig breed. Meat Science. 2013;**95**:658-666

[31] Papastergiadis A, Mubiru E, Van Langenhove H, De Meulenaer B. Malondialdehyde measurement in oxidized foods: Evaluation of the spectrophotometric thiobarbituric acid reactive substances (TBARs) test in various foods. Journal of Agricultural and Food Chemistry. 2012;**60**:9589-9594

[32] Grau A, Guardiola F, Boatella J, Barroeta A, Codony R. Measurement

of 2-thiobarbituric acid values in dark chicken meat through derivative spectrophotometry: Influence of various parameters. Journal of Agricultural and Food Chemistry. 2000;**48**:1155-1159

[33] Raharjo S, Sofos JN, Schmidt GR. Improved speed, specificity, and limit of determination of an aqueous acid extraction thiobarbituric acid-C18 method for measuring lipid peroxidation in beef. Journal of Agricultural and Food Chemistry. 1992;**40**:2182-2185

[34] Díaz P, Linares MB, Egea M, Auqui SM, Garrido MD. TBARs distillation method: Revision to minimize the interference from yellow pigments in meat products. Meat Science. 2014;**98**:569-573

[35] Jermann C, Koutchma T, Margas E, Leadley C, Ros-Polski V. Mapping trends in novel and emerging food processing technologies around the world. Innovative Food Science and Emerging. 2015;**31**:14-27

[36] Barba FJ, Grimi N, Vorobiev E. New approaches for the use of nonconventional cell disruption technologies to extract potential food additives and nutraceuticals from microalgae. Food Engineering Reviews. 2014;7:45-62

[37] Barbosa-Cánovas GV, Medina-Meza I, Candoğan K, Bermúdez-Aguirre D. Advanced retorting, microwave assisted thermal sterilization (MATS), and pressure assisted thermal sterilization (PATS) to process meat products. Meat Science. 2014;**98**:420-434

[38] Ros-Polski V, Koutchma T, Xue J, Defelice C, Balamurugan S. Effects of high hydrostatic pressure processing parameters and NaCl concentration on the physical properties, texture and quality of white chicken meat. Innovative Food Science & Emerging Technologies. 2015;**30**:31-52

[39] Campus M, Flores M, Martinez A, Toldra F. Effect of high pressure treatment on colour, microbial and chemical characteristics of dry cured loin. Meat Science. 2008;**80**:1174-1181

[40] He Z, Huang Y, Li H, Qin G, Wang T, Yang J. Effect of high-pressure treatment on the fatty acid composition of intramuscular lipid in pork. Meat Science. 2012;**90**:170-175

[41] Ma HJ, Ledward DA, Zamri AI, Frazier RA, Zhou GH. Effects of high pressure/thermal treatment on lipid oxidation in beef and chicken muscle. Food Chemistry. 2007;**104**:1575-1579

[42] Mariutti L, Orlien V, Bragagnolo N, Skibsted L. Effect of sage and garlic on lipid oxidation in high-pressure processed chicken meat. European Food Research and Technology. 2008;**227**:337-344

[43] McArdle RA, Marcos B, Kerry JP, Mullen A. Monitoring the effects of high pressure processing and temperature on selected beef quality attributes. Meat Science. 2010;**86**:629-634

[44] Medina-Meza LG, Barnaba C, Barbosa-Canovas GV. Effects of high pressure processing on lipid oxidation: A review. Innovative Food Science & Emerging Technologies. 2014;**22**:1-10

[45] Orlien V. High-pressure-meat processing and milk gels. High Pressure Processing. 2009;**12**:56-58

[46] Rivas-Canedo A, Juez-Ojeda C, Nunez M, Fernandez-Garcia E, et al. Food Chemistry. 2011;**124**:749-758

[47] Andres A, Adamsen C, Møller J, Ruiz J, Skibsted L. High-pressure treatment of dry-cured iberian ham. Effect on colour and oxidative stability during chill storage packed in modified atmosphere. European Food Research and Technology. 2006;**222**:486-491

[48] Bolumar T, Skibsted LH, Orlien V. Kinetics of the formation of radicals in meat during high pressure processing. Food Chemistry. 2012;**134**:2114-2120

[49] Schindler S, Krings U, Berger RG, Orlien V. Aroma development in high pressure treated beef and chicken meat compared to raw and heat treated. Meat Science. 2010;**86**:317-323

[50] Byrnea DV, Brediea WLP, Mottram DS, Martens M. Sensory and chemical investigations on the effect of oven cooking on warmed-over flavour development in chicken meat. Meat Science. 2002;**61**:127-139

[51] Tornberg E. Effects of heat on meat proteins. Implications on structure and quality of meat products. Meat Science. 2005;**70**:493-508

[52] Das AK, Rajkumar V. Effect of different fat level on microwave cooking properties of goat meat patties. Journal of Food Science and Technology. 2011;**50**:1-6

[53] Chandrasekaran S, Ramanathan S, Basak T. Microwave food processing—A review. Foodservice Research International. 2013;**52**:243-261

[54] Serrano A, Librelotto J, Cofrades S, Saánchez-Muniz FJ, Jiménez-Colmenero F. Composition and physicochemical characteristics of restructured beef steaks containing walnuts as affected by cooking method. Meat Science. 2007;**77**:304-313

[55] Dominguez R, Gomez M, Fonseca S, Lorenzo JM. Effect of different cooking methods on lipid oxidation and formation of volatile compounds in foal meat. Meat Science. 2014;**97**:223-230

[56] Abida J, Rayees B, Masoodi FA. Pulsed light technology: A novel method for food preservation. International Food Research Journal. 2014;**21**:839-848

[57] Falguera V, Pagán J, Garza S, Garvín A, Ibarz A. Ultraviolet processing of liquid food: A review. Part 1: Fundamental engineering aspects. Food Research International. 2011;**44**:1571-1579

[58] Guerrero-Beltrán JA, Barbosa-Cánovas GV. Advantages and limitations on processing foods by UV light. Food Science and Technology International. 2004;**10**:137-147

[59] Kim T, Silva JL, Chen TC. Effects of UV irradiation on selected pathogens in peptone water and on stainless steel and chicken meat. Journal of Food Protection. 2002;**65**:1142-1145

[60] Lyon SA, Fletcher DL, Berrang ME. Germicidal ultraviolet light to lower numbers of *Listeria monocytogenes* on broiler breast fillets. Poultry Science. 2007;**86**:964-967

[61] Koutchma T, Forney L, Moraru C. Principles and applications of UV technology. In: Koutchma T, Forney L, Moraru C, editors. Ultraviolet Light in Food Technology. Boca Raton, FL: CRC Press; 2009. pp. 1-32

[62] Raghavan GSV, Rennie TJ, Sunjka PS, Orsat V, Phaphuangwittayakul W, Terdtoon P. Overview of new techniques for drying biological materials with emphasis on energy aspects. Brazilian Journal of Chemical Engineering. 2005;**22**:195-201

[63] Freitas JA. Introdução à hygiene e conservação das matérias-primas de origem animal. 1st ed. São Paulo: Atheneu; 2015. p. 422

[64] Turp GY, Icier F, Kor G. Influence of infrared final cooking on color, texture and cooking characteristics of ohmically pre-cooked meatball. Meat Science. 2016;**114**:46-53

[65] Alam Khan K, Abrahem M. Effect of irradiation on quality of spices.

International Food Research Journal. 2010;**17**:825-836

[66] Otto C, Zahn S, Rost F, Zahn P, Jaros D, Rohm H. Physical methods for cleaning and disinfection of surfaces. Food Engineering Reviews. 2011;**3**:171-188

[67] Ahn DU, Lee EJ. Mechanisms and prevention of off-odor production and color changes in irradiated meat. In: Komolprasert V, Morehouse KM, editors. Irradiation of Food and Packaging: Recent Developments. Washington, DC: ACS Symposium Series; 2004. pp. 43-76

[68] Jo C, Ahn DU. Volatiles and oxidative changes in irradiated pork sausage with different fatty acid composition and tocopherol content. Journal of Food Science. 2000;**65**:270-275

[69] Zhu MJ, Mendonca A, Lee EJ, Ahn DU. Influence of irradiation and storage on the quality of ready-to-eat Turkey breast rolls. Poultry Science. 2004;**83**:1462-1466

[70] Kim IS, Jo C, Lee KH, Lee EJ, Ahn DU, Kang SN. Effects of low-level gamma irradiation on the characteristics of fermented pork sausage during storage. Radiation Physics and Chemistry. 2012;**81**:466-472

[71] Kang S, Park SY, Ha SD. Application of gamma irradiation for the reduction of norovirus in traditional Korean half-dried seafood products during storage. LWT- Food Science and Technology. 2016;**65**:739-745

[72] Contini C, Alvarez R, O'Sullivana M, Dowling DP, Gargan SO, Monahan FJ. Effect of an active packaging with citrus extract on lipid oxidation and sensory quality of cooked Turkey meat. Meat Science. 2014;**96**:1171-1176

[73] Pindi W, Mah HW, Munsu E, Ab Wahab N. Effects of addition of *Kappaphycus alvarezii* on physicochemical properties and lipid oxidation of mechanically deboned chicken meat (MDCM) sausages. British Food Journal. 2017;**119**:2229-2239

[74] Panda AK, Cherian G. Tissue tocopherol status, meat lipid stability, and serum lipids in broiler chickens fed *Artemisia annua*. European Journal of Lipid Science and Technology. 2017;**119**:1-7

[75] Bolumar T, Andersen ML, Orlier V. Antioxidant active packaging for chicken meat processed by high pressure treatment. Food Chemistry. 2011;**129**:1406-1412

[76] Badr HM, Mahmoud KA. Antioxidant activity of carrot juice in gamma irradiated beef sausage during refrigerated and frozen storage. Food Chemistry. 2011;**127**:1119-1130

Chapter 4

Antioxidants from Nigerian Medicinal Plants: What Are the Evidence?

Abubakar Babando Aliyu, Jonathan Ilemona Achika, Joseph Adesina Adewuyi, Patience Gangas, Hamisu Ibrahim and Adebayo Ojo Oyewale

Abstract

The search for natural antioxidants from plants would continue to be a dominant research interest for many years. This is because of the increasing understanding on the role of oxidative stress in damaging cell structures such as DNA, due to over production of free radicals and reactive oxygen species (ROS) in human systems, which are linked to inflammation, cancer and diabetes. However, phenolic compounds especially from phytochemicals or vegetable foods play important roles in reducing the risk of these diseases and reinforces the importance of natural antioxidants in human health. These antioxidant molecules neutralize or quench the ROS by either hydrogen atom transfer or single electron transfer mechanisms. Thus, the capacity to scavenge ROS and free radicals or inhibits lipid peroxidation is measured quantitatively as the strength of antioxidant activity. Several chemical and biochemical protocols have been used in the evaluation of plant extracts as antioxidants. Overwhelming literature reports have indicated varying degrees of antioxidant efficacies of extracts from Nigerian medicinal plants in comparison to synthetic antioxidants. These efficacies were analyzed to provide insight into the strength of antioxidant activity. This chapter reviewed 250 Nigerian medicinal plants in search of evidence for effective antioxidants.

Keywords: Nigerian medicinal plants, antioxidants, DPPH, ROS, free radicals

1. Introduction

Since the discovery of enzyme superoxide dismutase (SOD) and the evidence that emerged in support of the role of free radicals in biological systems, human understanding of free radical biochemistry in health and disease continue to advance [1]. This provided the basis for continuous search on natural antioxidants from foods and phytomedicines. Overwhelming reports on Google search engine has indicated 92,800 hits for "antioxidant activity" of medicinal plants in the last 10 years (2008–2018). This is due to growing interest on the antioxidant properties of medicinal plants. Several chemical and biochemical protocols have been used in the evaluation of antioxidant activity including the oxygen radical absorbance capacity (ORAC), total radical-trapping antioxidant potential (TRAP), total oxidant scavenging capacity (TOSC), chemiluminescence (CL), croton bleaching, low density lipoprotein (LDL) oxidation, ferric reducing antioxidant power (FRAP), copper reduction assay (CUPRAC), 2,2′ azinobis(3-ethylbenzothiazoline-6-sulfonic acid (ABTS) assay, 2,2-diphenyl-1-picrylhydrazyl (DPPH), nitric oxide (NO), hydroxyl radical (OH) hydrogen peroxide (H_2O_2) and total phenolic assay among others [2]. Biochemical protocols are based on animal models for *in vivo* evaluations of oxidative stress biomarkers. However, this study is focused on *in vitro* evaluations of antioxidants from plants based on hydrogen atom transfer or single electron transfer mechanisms [2]. The strength of antioxidant activity measured from a combination of different methodologies was used to evaluating antioxidant effectiveness [3]. This review provides fundamental background on free radical and ROS in human health and disease with a view to understand the roles of natural antioxidants. We reviewed 250 Nigerian medicinal plants evaluated for antioxidant activity in search of evidence for effective antioxidants.

2. Reactive oxygen species (ROS) in human health and disease

Human system uses oxidation for normal metabolic activities in the transformation of nutrients into energy. During oxidation, reactive oxygen species (ROS) are also produced at low levels in normal physiological conditions, which are necessary for maintaining normal cell functions such as signaling immunity and homeostasis [4]. These activities are maintained by endogenous antioxidant (enzymatic) defense systems produced by the body for protection against harmful effects. These include superoxide dismutase (SOD), glutathione peroxidase (GSH-Px), glutathione reductase (GSH-Rx) and catalase [5]. Excessive production of ROS beyond the body defense mechanisms can be extremely harmful to cellular functions by damaging nucleic acids, oxidizing proteins, and causing lipid peroxidation [6]. The resultant cell damage by free radicals and ROS appeared as major contributor to aging and degenerative diseases of aging such as cancer, cardiovascular disease, immune system decline, liver diseases, diabetes mellitus, inflammation and brain dysfunction among others [7, 8]. These ROS and reactive nitrogen species (RNS) including superoxide anion O_2^{-*}, hydroxide ion OH^{-*}, hydroxyl radical OH^*, peroxyl radical ROO^* and nitric oxide NO^* as well as H_2O_2, lipid peroxides ROOH, and singlet O_2 are very reactive and can initiates free radical reactions or lipid peroxidation in living cells.

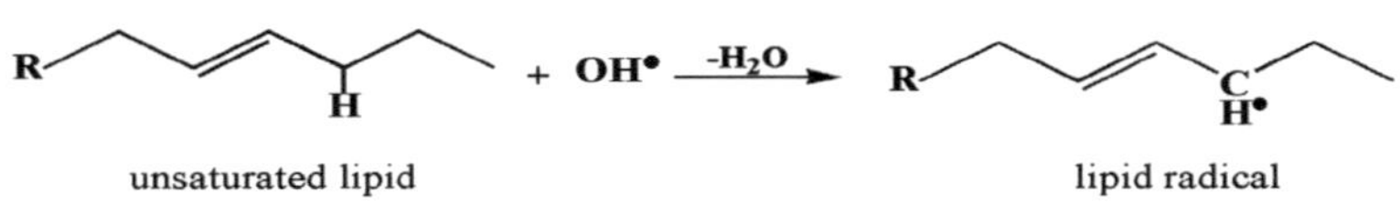

ROS can be produced either by external sources (e.g., tobacco smoke, stress, etc.), as by-products during the mitochondrial electron transport of aerobic respiration or by oxidoreductase enzymes and metal catalyzed oxidation [9]. But the biological effects of ROS depend on the types of cell or tissue in relation to enzyme production, signal transduction and DNA repair [10]. ROS are harmful when excessive productions are not balanced by body antioxidant mechanism. This imbalance between ROS production and enzymatic antioxidant defense systems is called oxidative stress [11]. Antioxidants counteract oxidative stress by neutralizing free radicals because they are reducing agents that react with and buffer ROS as a form of defense against oxidative stress [12].

3. Phytochemicals as sources of natural antioxidants

Antioxidants are molecules that prevent oxidation or inactivates the reactive oxygen species and thus prevent oxidative damage to the cells and body tissues [13]. Antioxidants can also inhibit, quench or scavenge free radicals converting them into new and stable chemical compounds [14]. Broadly, antioxidants are classified as enzymatic and non-enzymatic with each class providing complementary role of protection against free radicals in human systems. Previous work has concisely discussed on antioxidants classification [3] as summarily reproduced in **Figure 1**. But our focus is the non-enzymatic antioxidant involving flavonoids, phenolic acids, vitamins, carotenoids, minerals and cofactors. They are exogenous sources of protection through diet. Plants foods contain a variety of nutrients and non-nutrients chemicals which are good antioxidant agents. These sources of natural antioxidants including Vitamin A (retinol) obtained from β-carotene, vitamin C (ascorbic acid), Vitamin E (α-tocopherol), lycopene and carotenoids occur naturally in fruits, vegetables, legumes and grains which are commonly consumed and play important role in the defense against free radicals [3, 15]. Medicinal plants are rich source of phenolic compounds such as flavonoids, phenolic acids and coumarins [16, 17]. Flavonoids are antioxidants compounds composed of anthocyanins, flavanones, flavonols, flavones, isoflavonoids and flavanones, while hydroxycinnamic and hydroxybenzoic acids such as gallic acid are components of phenolic acids widely distributed secondary metabolites in plants with antioxidant and antiradical properties [18]. They are important as chelators and free radical scavengers of hydroxyl and peroxyl radicals, superoxide anions and peroxynitrites [19]. Carotenoids natural pigments are important phytochemical antioxidants obtained from plants. They are structurally grouped into carotenes and xanthophyll based on the degree of oxygenation of carotenoid hydrocarbons and exert antioxidant effect by singlet oxygen quenching ability [3]. Several studies on the antioxidant activities of various herbal plants have indicated their enormous medicinal values as inhibitors of free radical and ROS [20].

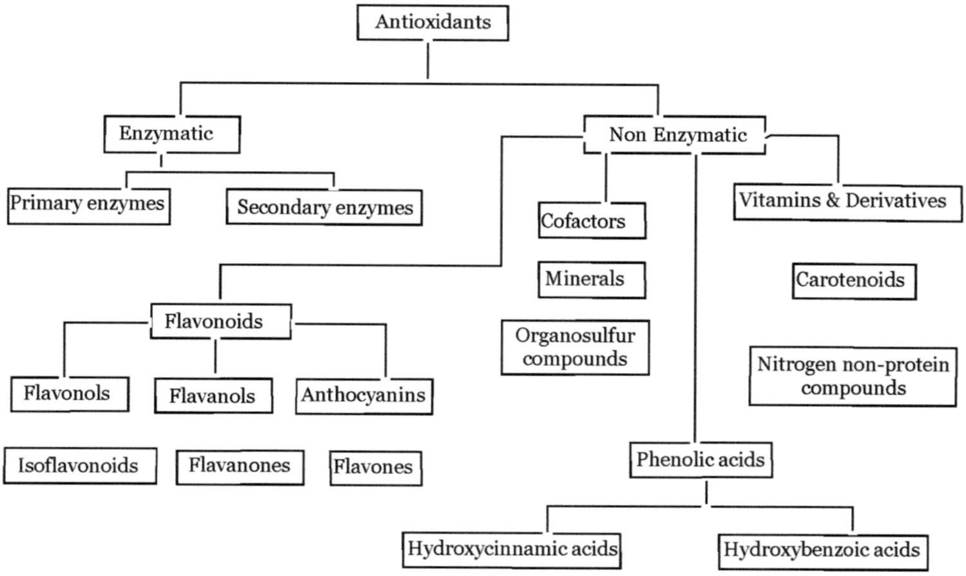

Figure 1.
Board classification of antioxidants adapted from Carocho and Ferreira [3].

4. Antioxidants from Nigerian medicinal plants

Nigeria is a west African country with an area of 923,769 km^2 having a population of 198 million with 250 ethnic groups [21]. The country shares border with republic of Cameroun to the east, Niger and Chad republics to the north, Benin republic to the west and Gulf of Guinea to the south. Nigeria has favorable climate conditions with enormous diversity of plant species, which are distributed across geographical contrast of the mangrove swamps in South-South (SS), to the tropical rain forests covering South-West (SW) and South-East (SE) and to the grassland vegetation of North-Central (NC) up to the Sahel savannah of semi-arid North-East (NE) (**Figure 2**). Many of these plants are used as medicines for treatment of illness or management of human and animal health among rural and urban dwellers.

The application of herbal recipes especially in the management of human metabolic diseases such as diabetes and cancer is common knowledge among Nigerians. This prompted research interest in academia on the potentials of phytomedicines as complimentary or alternative treatment agents, and consequent research efforts to validate their pharmacological properties. The number of Nigerian medicinal plants reported for antioxidants is enormous. However, 250 medicinal plants evaluated for antioxidant activity were studied in addition to the 28 compounds isolated from 44 plants. But antioxidant evaluations on crude extracts rather than on pure compounds largely dominated the literature. Thus, effective activity based on concentrations required to inhibit 50% free radicals (IC_{50}) for selected extracts are presented (**Table 1**) together with concentrations of various standard antioxidants used.

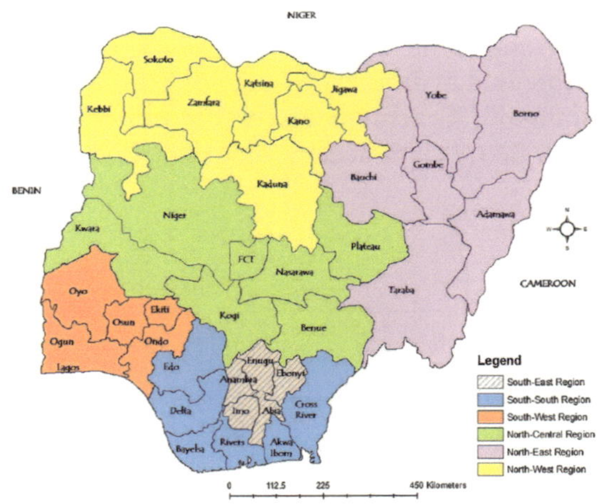

Figure 2.
Map of Nigeria showing the six geopolitical zones.

S. No	Plant name	Family	Part used	IC_{50} sample	IC_{50} standard	Ref.
1	*Abrus precatorius*	*Leguminosae*	Seed	1.92* 2.10*	AA = 1.83 AA = 1.20	[37]
2	*Acalypha ornata*	Euphorbiaceae	Leaf	20.50*	TC = 15.4	[68]
3	*Acalypha wilkesiana*	Euphorbiaceae	Leaf	15.25*	AA = 7.26	[26]
4	*Acanthospermum hispidum*	Asteraceae	Aerial	28.9*	AA = 1.41	[39]
5	*Aframomum melegueta*	Zingiberaceae	Fruit Leaf	0.04** 0.07**	AA = 0.03	[69]

S. No	Plant name	Family	Part used	IC_{50} sample	IC_{50} standard	Ref.
6	*Ageratum conyzoides*	Asteraceae	Leaf	31.25*	AA = 7.26	[26]
7	*Allamanda cathartica*	Apocynaceae	Leaf	0.46**	VE = 0.25	[70]
8	*Allanblackia floribunda*	Guttiferae	Leaf	0.02** 0.1**	VE = 0.01	[71]
9	*Alstonia boonei*	Apocynaceae	Stem	0.12**	AA = 0.06	[72]
10	*Alstonia congensis*	Apocynaceae	Root	19.7*	AA = 4.9	[73]
11	*Alternanthera dentata*	Amaranthaceae	Leaf	35*	AA = 125	[46]
12	*Amaranth caudatus*	Amaranthaceae	Leaf Stem	15.81**	TC = 13.2	[27]
13	*Anacardium occidentale*	Anacardiaceae	Bark Leaf	5.66* 7.77*	AA = 4.57	[74]
14	*Annona senegalensis*	Annonaceae	Leaf	45.72* 49.0*	GA = 48.77 TX = 72.9	[35]
15	*Aspilia africana*	Asteraceae	Leaf	160*	AA = 120	[75]
16	*Asystasia gangetica*	Acanthaceae	Leaf	100*	AA = 150	[75]
17	*Bauhinia galpinii*	Caesalpiniaceae	Leaf	20.52*	AA = 19.8	[76]
18	*Bauhinia monandra*	Caesalpiniaceae	Leaf	5.56*	AA = 30.0	[45]
19	*Bixa orellana*	Bixaceae	Leaf	0.45*	VE = 0.25	[70]
20	*Borreria ocymoides*	Rubiaceae	Aerial	1.85**	AA = 0.05	[77]
21	*Borreria verticillata*	Rubiaceae	Leaf	2.98*	AA = 1.05	[78]
22	*Bridelia ferruginea*	Euphorbiaceae	Leaf	12.5*	AA = 7.26	[26]
23	*Bridelia micrantha*	Euphorbiaceae	Leaf	0.1 μM	AA = 2.0 μM	[38]
24	*Bryophyllum pinnatum*	Crassulaceae	Leaf	0.41**	VC = 0.067	[79]
25	*Calliandra surinamensis*	Mimosaeae	Flower	28**	VE = 38	[80]
26	*Calyptrochilum christyanum*	Orchidaceae	Whole	50.6*	AA = 1.41	[39]
27	*Canthium subcordatum*	Rubiaceae	Leaf	23.9*	AA = 4.9	[74]
28	*Capsicum annuum*	Solanaceae	Fruit	1.15**	BHA = 0.96	[25]
29	*Capsicum frutescens*	Solanaceae	Fruit	0.67**	BHA = 0.96	[25]
30	*Carica papaya*	Caricaceae	Seed	0.227**	AA = 0.109	[67]
31	*Cassia sieberiana*	Leguminosae	Leaf	24.1* 46.6*	AA = 4.9 TC = 38.9	[73]
32	*Cassia singuaena*	Leguminosae	Leaf	1.20*	AA = 2.56	[48]
33	*Celosia trigyna*	Amaranthaceae	Leaf	120*	AA = 120	[75]
34	*Cissampelos owariensis*	Menispermaceae	Leaf	2.77*	AA = 0.067	[25]
35	*Citrus aurantifolia*	Rutaceae	Peel	12.1**	VC = 0.067	[25]
36	*Commiphora kerstingii*	Burseraceae	Leaf	0.33** 0.54**	AA = 0.49	[23]
37	*Corchorus olitorius*	Malvaceae	Leaf	11.8** 27.52*	TC = 13.2 AA = 188.3	[27] [33]
38	*Crassocephalum rubens*	Asteraceae	Leaf	2.91** 1.73**	VC = 1.18 VC = 0.56	[62]
39	*Cucumis sativus*	Cucurbitaceae	Leaf	1.68** 71.1**	BHA = 0.96 TC = 13.2	[27] [25]

S. No	Plant name	Family	Part used	IC_{50} sample	IC_{50} standard	Ref.
40	*Cucurbita moschata*	Cucurbitaceae	Leaf	150*	AA = 120 TC = 50	[75]
41	*Cymbopogon citratus*	Poaceae	Leaf	1.35*	VE = 0.25	[70]
42	*Daniellia oliveri*	Leguminosae	Leaf	15.5*	TC = 0.25	[50]
43	*Daucus carota*	Apiaceae	Aerial	4.61**	BHA = 0.96	[25]
44	*Ehretia cymosa*	Boraginaceae	Leaf	0.47**	GA = 2.09	[53]
45	*Emilia coccinea*	Asteraceae	Leaf	120*	AA = 120	[75]
46	*Eugenia caryophyllata*	Myrtaceae	Leaf Bud	0.03* 0.02*	AA = 0.03	[47]
47	*Eupatorium adenophorum*	Asteraceae	Root	22.4* 53.7*	AA = 4.9 RT = 3.3	[73]
48	*Eupatorium odoratum*	Asteraceae	Leaf	0.07**	AA = 0.06	[72]
49	*Euphorbia hirta*	Euphorbiaceae	Leaf	2.5**	VC = 4.5	[81]
50	*Feretia apodanthera*	Rubiaceae	Root	0.053**	VC = 0.048	[43]
51	*Ficus exasperata*	Moraceae	Leaf	0.86*	VE = 0.25	[70]
52	*Ficus gnaphalocarpa*	Moraceae	Leaf	45.3* 44.6*	GA = 48.8 TX = 72.9	[34]
53	*Ficus sycomorus*	Moraceae	Stem	42.0*	VC = 25.0	[82]
54	*Globimetula oreophila*	Loranthaceae	Leaf	0.38**	VC = 0.06	[79]
55	*Gongronema latifolia*	Asclepiadaceae	Leaf	70.0*	VC = 50	[83]
56	*Grewia carpinifolia*	Tiliaceae	Leaf Stem	0.32** 0.39**	AA = 0.31 AA = 0.18	[32]
57	*Harungana madagascariensis*	Hypericaceae	Stem	37.5*	BHT = 16.2	[36]
58	*Heliotropium indicum*	Boraginaceae	Aerial	48.4*	AA = 1.41	[39]
59	*Hibiscus sabdariffa*	Malvaceae	Leaf	0.14*	AA = 0.02	[84]
60	*Holarrhena floribunda*	Apocynaceae	Leaf	7.2*	QT = 2.95	[85]
61	*Ipomoea asarifolia*	Convulvulaceae	Leaf	24.3*	AA = 1.41	[39]
62	*Irvingia gabonensis*	Irvingiaceae	Root Stem	12.4* 25.5*	AA = 4.9 TC = 38.9	[73]
63	*Justicia secunda*	Acanthaceae	Leaf	1.58 μM	AA = 2.52	[86]
64	*Kalanchoe pinnata*	Crassulaceae	Leaf	180*	AA = 120	[75]
65	*Lactuca sativa*	Asteraceae	Whole	0.26**	QT = 0.83	[25]
66	*Landolphia owariensis*	Apocynaceae	Root	8.8* 49.1*	AA = 4.9 TC = 38.9	[73]
67	*Laportea ovalifolia*	Urticaceae	Leaf	100*	AA = 150	[75]
68	*Lasianthera africana*	Icacinaceae	Leaf Root	0.30** 0.27**	RT = 0.26	[28]
69	*Launaea taraxacifolia*	Asteraceae	Shoot Leaf	1.94** 1.59**	VC = 1.18 VC = 0.56	[62]
70	*Lawsonia inermis*	Lythraceae	Leaf	3.80*	AA = 7.26	[26]
71	*Leptadenia hastata*	Asclepiadaceae	Leaf	42.3*	GA = 48.8	[35]
72	*Lycopersicon esculentum*	Solanaceae	Fruit	1.16** 1.47**	QT = 0.83	[25]

S. No	Plant name	Family	Part used	IC_{50} sample	IC_{50} standard	Ref.
73	*Massularia acuminata*	Rubiaceae	Leaf	70.0*	VC = 7.59	[87]
74	*Mondia whitei*	Apocynaceae	Leaf	6.1*	AA = 3.4	[66]
75	*Moringa oleifera*	Moringaceae	Leaf	0.16*	AA = 0.02	[84]
76	*Murraya koenigii*	Rutaceae	Leaf	7.35*	TC = 13.2	[27]
77	*Nauclea diderrichii*	Rubiaceae	Stem	18.12*	AA = 1.41	[39]
78	*Nauclea latifolia*	Rubiaceae	Leaf	12.9*	AA = 4.9	[73]
79	*Ocimum basilicum*	Lamiaceae	Leaf	1.0*	AA = 9.0	[55]
80	*Ocimum gratissimum*	Lamiaceae	Leaf Stem	0.14* 8.67*	AA = 0.02 BHA = 3.36	[33] [27]
81	*Parinari curatellifolia*	Chrysobalanaceae	Leaf	13.5*	VC = 1.98	[88]
82	*Parkia biglobosa*	Leguminosae	Stem	15.65*	AA = 7.26	[26]
83	*Phragmanthera capitata*	Loranthaceae	Leaf	1.9* 1.0*	BHT = 4.6 VC = 10	[41]
84	*Piliostigma reticulatum*	Fabaceae	Leaf	10.3*	AA = 3.9	[22]
85	*Piliostigma thonningii*	Fabaceae	Leaf	14.7*	AA = 3.9	[22]
86	*Piper guineense*	Piperaceae	Seed	74*	AA = 31.7	[89]
87	*Psidium guajava*	Myrtaceae	Leaf	0.04**	BHA = 0.05	[24]
88	*Sapium ellipticum*	Euphorbiaceae	Stem	0.19**	BHT = 0.11	[90]
89	*Senna alata*	Fabaceae	Leaf	0.59**	VC = 0.067	[79]
90	*Simarouba glauca*	Simaroubaceae	Stem	4.7*	BHT = 5.0	[42]
91	*Solanum macrocarpon*	Solanaceae	Leaf	6.21**	TC = 13.2	[27]
92	*Spinacia oleracea*	Amaranthaceae	Leaf	12.6*	TC = 13.2	[27]
93	*Spondias purpurea*	Anacardiaceae	Stem	8.3*	AA = 11.5	[52]
94	*Stachytarpheta jamaicensis*	Verbenaceae	Leaf	5.0*	AA = 9.0	[51]
95	*Strophanthus hispidus*	Apocynaceae	Root	1.18**	VC = 0.067	[79]
96	*Telfairia occidentalis*	Cucurbitaceae	Leaf	0.16**	AA = 0.02	[84]
97	*Trichilia catigua*	Meliaceae	Stem	30.28*	AA = 20.72	[64]
98	*Vernonia amygdalina*	Asteraceae	Leaf	31.25*	AA = 7.26	[26]
99	*Vernonia calvoana*	Asteraceae	Leaf	1.90 μM	AA = 2.0 μM	[49]
100	*Vernonia cinerea*	Asteraceae	Leaf	6.50* 8.0*	GA = 0.62	[30]
101	*Vernonia migeodii*	Asteraceae	Leaf	20.0*	AA = 18.0	[91]
102	*Vitex doniana*	Verbenaceae	Leaf	53.23*	GA = 48.8	[34]
103	*Xylopia aethiopica*	Annonaceae	Fruit	1.04**	VC = 0.067	[79]
104	*Zingiber officinale*	Zingiberaceae	Rhizome	47.0*	AA = 36.4	[92]

AA, ascorbic acid; QT, quercetin; RT, rutin; GA, gallic acid; VC, vitamin C; VE, vitamin E; TX, trolox; BHT, butylated hydroxy toluene ; BHA, butylated hydroxy anisole; TC, tocopherol.
*$IC_{50} = \mu gmL^{-1}$.
**$IC_{50} = mgmL^{-1}$.

Table 1.
Antioxidant activities of selected Nigerian plants.

5. Antioxidant activities of crude extracts

The antioxidant efficacies of Nigerian plants were largely evaluated using protocols involving DPPH, ABTS, FRAP, TAC, NO, OH and or H_2O_2 targets. The DPPH radical scavenging assay is one of the commonly used techniques for quick evaluation of antioxidant capacity. Plant extracts tested for DPPH inhibition have demonstrated interesting efficacies for instance, crude extracts of *P. reticulatum* (40.10 μgmL^{-1}) and *P. thoninngii* (50.94 μgmL^{-1}) showed comparable activity with *Ginkgo biloba* (EC_{50} 40.72 μgmL^{-1}) [22]. Nigerian plants evaluated for antioxidants activity between 2008 and 2012 were reported in 40 publications representing over 166 extracts from 119 plants. These studies showed 29 extracts with effective activity on various free radical targets. However, 15 extracts have comparable antioxidant efficacies to standard antioxidants, while 14 have higher percent (%) inhibition or lower IC_{50} values than the standards used. These include stem methanol extract of *C. kerstingii* (IC_{50} 26.27 μgmL^{-1}, ascorbic acid 33.59 μgmL^{-1}) [23] and leaf methanol extract of *P. guajava* (IC_{50} 0.037 $mgmL^{-1}$, BHA 0.049 $mgmL^{-1}$) [24]. But DPPH inhibition studies on selected vegetable plants showed better effective activity for *L. sativa* (IC_{50} 0.26 $mgmL^{-1}$), *Z. officinale* (IC_{50} 0.29 $mgmL^{-1}$) and *C. frutescens* (IC_{50} 0.67 $mgmL^{-1}$) respectively compared to BHA (IC_{50} 0.96 $mgmL^{-1}$) and quercetin (IC_{50} 0.83 $mgmL^{-1}$) [25]. The activity of *L. inermis* was most profound of the 36 medicinal plants surveyed in Southwestern Nigeria, with lower IC_{50} of 3.80 μgmL^{-1} than ascorbic acid (7.26 μgmL^{-1}) [26]. Similar evaluations of DPPH inhibition on 15 medicinal plants showed *S. oleracea* extract with lower IC_{50} of 12.6 $mgmL^{-1}$. But *S. macrocarpon* extract was most effective with IC_{50} 6.21 $mgmL^{-1}$ lower than α-tocopherol (13.20 $mgmL^{-1}$) [27].

The analysis of antioxidant efficacies on medicinal plants reported from 2013 to 2017 in 55 publications, involving 211 extracts from 144 plants was carried out. We observed that 70 extracts from 50 plants have exhibited good antioxidant efficacies on various free radical targets with 51 extracts from 53 plants having comparable efficacies to standard antioxidants. However, lower IC_{50} or higher percent (%) inhibitions compared to standards were observed with 20 extracts from 17 medicinal plants. The $NO^{\bullet}$ inhibition on root extract of *L. africana* (IC_{50} 0.27 $mgmL^{-1}$) compared very well with rutin (IC_{50} 0.28 $mgmL^{-1}$) [28]. The DPPH inhibition on *P. guajava* (IC_{50} 1.564 μgmL^{-1}) extract also indicated effective activity compared to ascorbic acid (IC_{50} 5.950 μgmL^{-1}) [29]. Other plant extracts including *V. cinerea* (IC_{50} 6.50 μgmL^{-1}) compared to gallic acid (IC_{50} 0.62 μgmL^{-1}) [30] and *K. senegalensis* stem bark (IC_{50} 95.76 μgmL^{-1}) with ascorbic acid (223.35 μgmL^{-1}) indicated effective activity [31]. The inhibition of lipid peroxidation using thiobarbituric acid reactive substances (TBARS) assay on leaf extract of *G. carpinifolia* was very effective with IC_{50} of 0.24 $mgmL^{-1}$ compared to ascorbic acid (IC_{50} 0.38 $mgmL^{-1}$). Moreover, the ABTS assay indicated 100% inhibitions for both extracts and ascorbic acid [32]. The antioxidant evaluations on two of the most locally utilized vegetable plants such as *V. amygdalina* and *O. gratissimum* showed effective inhibitions compared to the standard ascorbic acid [33].

Furthermore, DPPH inhibitions on *S. occidentalis* (IC_{50} 42.80 μgmL^{-1}) compared to gallic acid (48.77 μgmL^{-1}) was effective, but ABTS assay on *F. gnaphalocarpa* (44.63 μgmL^{-1}) was more effective than Trolox (72.92 μgmL^{-1}) [34]. Similarly, *L. hastata* (IC_{50} 42.32 μgmL^{-1}) when compared to gallic acid (48.77 μgmL^{-1}) and ABTS on *A. senegalensis* (IC_{50} 48.98 μgmL^{-1}) with Trolox (72.92 μgmL^{-1}) have interesting lower IC_{50}values [35]. But *H. madagascariensis* exhibited moderate activity [36] while *A. precatorius* [37] and *B. micrantha* [38] have demonstrated effective inhibitions (**Table 1**). The analyses of Nigerian plants in 2018 showed interesting activities with 15 plant extracts from 32 published reports. Plants with moderate DPPH inhibition include *A. hispidum, A. laxiflora, C. christyanum, H. indicum* and

I. asarifolia [39]. However, effective inhibitions were observed on root extracts of *D. tripetala* (IC_{50} 0.631 μgmL^{-1}) and *M. excelsa* (IC_{50} 0.194 μgmL^{-1}) compared to 4.60 μgmL^{-1} ascorbic acid [40]. Similarly, *P. capitata* (27.4 μgmL^{-1}) was effective than BHT (56.0 μgmL^{-1}) [41], and the evaluation of *S. glauca* stem bark on FRAP (4.70 μgmL^{-1}) and $NO^{\bullet}$ (11.90 μgmL^{-1}) were effective than 5.0 μgmL^{-1} and 18.0 μgmL^{-1} of BHT respectively [42]. Lastly, plant crude extracts have demonstrated varying but strong efficacies on different free radical targets which in many cases surpassed standard antioxidants. The report on DPPH inhibition of *F. apodanthera* root bark ethanol extract represents effective activity with IC_{50} of 0.053 μgmL^{-1} in comparison to vitamin C (0.048 μgmL^{-1}) standard [43].

6. Chemical composition and antioxidant activity

6.1 GC-MS analysis of extracts and evaluation of antioxidant activity

The antioxidant evaluations of Nigerian medicinal plants with determination of chemical composition using gas chromatography-mass spectrometry (GC-MS) have become routine studies. The GC-MS is intended to give insight on the probable chemical entities of volatile components present in the sample extract. Several plants constituents have been analyzed using GC–MS by comparison of compounds' retention times with library of standard chemical entities provided by the National Institute of Standards and Technology (NIST) database imbedded in the instrument. The chemical constituents with low molecular weights such as terpenoids, long chain alkanes, phenolics and fatty acid methyl esters (FAME) are separated and detected by GC-MS. This is perhaps one reason that FAME are prevalent from among plant extracts, but sharp contrast between lipophilic and hydrophilic components are determined by solvent polarity or method of extraction [44].

The GC-MS analyses and evaluation of antioxidants on *B. monandra* hexane extract showed 4-hydroxy-5-methyl-3-propyl-2-hexanone (42.7%) and oleic acid (20%) as major compounds. The DPPH inhibition (IC_{50} 5.56 μgmL^{-1}) with ascorbic acid (IC_{50} 30.0 μgmL^{-1}) showed interesting efficacy, but ethyl acetate extract containing largely oleic acid (40.76%) and hexadecanoic acid (21.75%) was more effective (IC_{50} 0.01 μgmL^{-1}) [45]. The evaluation on *A. dentata* methanol extract containing hexadecanoic acid (31.6%), phytol (24.6%) and octadecanoic acid (10.56%) was found to be poor. However, the FRAP inhibition showed optimum activity (0.65 $\mu molL^{-1}$) compared to ascorbic acid (2.00 $\mu molL^{-1}$) [46]. The DPPH screening on buds, leaf, root and stem of commonly used spice, *E. caryophyllata* was reported. The various ethanol extracts showed effective activities IC_{50} of 0.02, 0.03, 3.66 and 0.99 μgmL^{-1} respectively, compared to ascorbic acid (IC_{50} 0.03 μgmL^{-1}) and gallic acid (IC_{50} 0.05 μgmL^{-1}) standards. This indicated an important response to the DPPH scavenging capacity which have been largely attributed to aromatic phenols, caryophyllene, aromatic esters and ethers [47]. Similar comprehensive study on the leaf, stem bark and root of *C. singueana* was reported. The DPPH, OH and $NO^{\bullet}$ showed IC_{50} of 1.20, 2.58 and 35.99 μgmL^{-1} for DPPH inhibition of stem bark ethanol, root aqueous and leaf ethanol extracts respectively. But the response on OH showed IC_{50} of 1.58, 2.05 and 6.47 μgmL^{-1} respectively, for stem bark ethyl acetate, root aqueous and leaf ethanol extracts. The $NO^{\bullet}$ results however, was interesting on leaf aqueous extract (IC_{50} 2.81 μgmL^{-1}) better than the ascorbic acid (IC_{50} 26.28 μgmL^{-1}) and Trolox (IC_{50} 599.21 μgmL^{-1}) standards used. The chemical components such as resorcinol (54%) and phytol (23.7%) were largely detected from ethanol extracts of stem bark and leaf respectively [48].

The leaf ethyl acetate extract of *V. calvoana* harvested from the South–South Nigeria contains largely aromatic compounds such as ethyl benzene (22%) and 1,2,3-trimethyl benzene (12.5%). FRAP inhibitions on extract (1.98 μM) and ascorbic acid (2.0 μM) were more effective than on DPPH [49]. Although the inhibition on DPPH by plant extracts have been promising but chloroform extract of *D. oliveri* exudate showed rather poor (IC_{50} of 15.5 μgmL^{-1}) when compared to α-Tocopherol (0.25 μgmL^{-1}) [50]. But *S. jamaicensis* methanol extract (IC_{50} 5.0 μgmL^{-1}) was more effective than ascorbic acid (IC_{50} 9.0 μgmL^{-1}). Compounds such as 3,5-dihydroxy-6-methyl-2,3-dihydro-4H-pyran-4-one (13.7%) and D-arabinitol (13.5%) have been largely identified [51]. It was interesting to note that *S. purpurea* hexane extract showed effective DPPH inhibition (IC_{50} 8.3 μgmL^{-1}) than ascorbic acid (IC_{50} 11.5 μgmL^{-1}) [52]. The evaluation of *E. cymosa* leaf extracts on ethyl acetate (IC_{50} 0.56 $mgmL^{-1}$) and methanol (IC_{50} 0.60 $mgmL^{-1}$) extracts justifies effective activity compared to gallic acid (IC_{50} 0.47 $mgmL^{-1}$). The 2-hexadecycloxirane (34.2%) and methyl linoleate (28.9%) were detected as major components of methanol extract [53]. Lastly, the GC-MS analyses of various plant extracts and antioxidant evaluation have revealed similar pattern of contents and composition in addition to contrasting influences of solvent polarity to radical inhibition efficacies. Nevertheless, interesting antioxidant efficacies were observed.

6.2 GC-MS analysis of essential oils and evaluation of antioxidant activity

The essential oils (EO) from Nigerian medicinal plants have been analyzed using the GC–MS and evaluated for antioxidants activity. Because they are mixtures of several constituents containing largely low molecular weights compounds, EO are rapidly analyzed using GC–MS to ascertain their chemical composition. The essential oils (EO) from *P. guajava* showed that 3, 6-dioxa-2,4,5,7-tetraoctane-2,2,4,4,5,5,7,7-octamethyl (11.7%) and cyclononane (10.7%) are largely identified. DPPH inhibition showed 71.83% comparable to 68.7% of ascorbic acid [54]. The antioxidant efficacy of *O. basilicum* EO was also interesting (IC_{50} of 1.0 μgmL^{-1}) probably due to phenolic constituents such as methyl eugenol (15.5%), o-nitrocumene (14.0%) and 2-phenyl-1-hexanol (14.0%) [55]. The EO of common spices such as *A. melegueta* leaf, *C. crepidioides* stem bark, *O. gratissimum* leaf and *M. myristica* stem bark showed various chemical constituents with interesting antioxidant activity. The EO components largely identified from the four plants are myrtenyl acetate (29.1%), thymol (44%), γ-terpinene (53%) and γ-cardinene (31.1%) respectively. The highest antioxidant activity was found from EO of *O. gratissimum* (96.4%) which compared to BHA (96.7%). However, other EOs have demonstrated radical scavenging of >50% inhibition [56]. The leaf EO of *C. portoricensis* was found to contain thymol (9.64%) and β-caryophyllene (9.15%) as main compounds which might have resulted to 75% DPPH inhibition compared to BHT (95%) [57]. Similarly, analysis on *M. alternifolius* EO that yielded largely tricosane (19.45%) and z-14-nonanosene (13.4%) with interesting efficacy (97.95%) compared to ascorbic acid (97.88%) [58]. This trend of activity demonstrated by the EO was observed in *E. maculata* which contained α-pinene (8.0%), β-trans-ocimene (8.0%), 1S-α-pinene (7.0%) and cyclofenchene (7.0%) as main components, with the DPPH (IC_{50} 8.0 μgmL^{-1}) and FRAP (10.0 μgmL^{-1}) inhibition efficacies in comparison to 9.0 and 20.0 μgmL^{-1} ascorbic acid respectively [59]. Lastly, GC-MS analyses of EO from Nigerian plants have revealed interesting but similar chemical compounds with some degree of antioxidant efficacies. EO composition containing phenolics moieties and terpenoids have indicated evidence of effective radical inhibitions.

6.3 HPLC analysis of extracts and evaluation of antioxidant activity

The high-performance liquid chromatography (HPLC) has been reported in the analysis of major chemical constituents of plant extracts alongside with the antioxidant activity. The HPLC technique uses reverse phase chromatography because of simplicity, versatility and sensitivity towards separation, purification, quantification and identification of diverse natural products such as plant phenolics, steroids, alkaloids and flavonoids [60]. Hence, the combination of HPLC methods with antioxidants evaluations may provide the needed understanding of antioxidant efficacies of plant extracts. Previous HPLC profiling of ethanol extract of *Z. zanthoxyloide* showed quercetin, kaempferol and caffeic acid largely quantified. DPPH inhibition (IC_{50} 38.58 μgmL^{-1}) in comparison to ascorbic acid (6.63 μgmL^{-1}) was poor [61]. Similarly, aqueous extracts of *L. taraxacifolia* (IC_{50} 6.59 μgmL^{-1}) and *C. rubens* (IC_{50} 6.21 μgmL^{-1}) were less effective than Trolox (IC_{50} 0.51 μgmL^{-1}). Although methanol extracts of both plants contain gallic acid, caffeic acid, quercetin, rutin, isoquercetin and kaempferol as the main compounds identified, yet the activity was not interesting. But the OH inhibition on aqueous extracts showed rather interesting results with IC_{50} 1.94 and 1.09 μgmL^{-1} in comparison to IC_{50} 1.18 μgmL^{-1} of vitamin C [62].

Although antioxidant activities of plant extracts using DPPH have been established to correlate with phenolics and flavonoids contents [63]. However, many of the plants evaluated for antioxidants activity have no correlation with the number and amounts of phenolics and flavonoids quantified by HPLC. The report on *T. catigua* ethanol, ethyl acetate, dichloromethane and butanol extracts showed DPPH inhibition with IC_{50} of 9.17, 30.28, 42.42 and 76.35 μgmL^{-1} respectively. These, in comparison to ascorbic acid (20.72 μgmL^{-1}) indicated poor activity except the ethanol extract with lower IC_{50} than the standard. The extract was quantified to be rich in gallic acid, chlorogenic acid rutin and quercetin [64]. Similarly, *S. dulcificum* contains phenolic acids and flavonoids but demonstrated poor efficacies on DPPH (IC_{50} 139.45 μgmL^{-1}), ABTS (IC_{50} 135.83 μgmL^{-1}), $NO^{\bullet}$ (IC_{50} 119.17 μgmL^{-1}) and OH (IC_{50} 147.65 μgmL^{-1}) [65]. However, *M. whitei* contains largely caffeic acid with interesting efficacies on $NO^{\bullet}$ (IC_{50} 6.1 μgmL^{-1}) and FRAP (IC_{50} 5.7 μgmL^{-1}) compared to ascorbic acid (3.4 and 7.0 μgmL^{-1}) respectively [66]. Similarly, *C. papaya* seeds protein analyzed using the LC-ESI-DAD-MS with largely ferulic acid in addition to flavonoid sugars, justifies the antioxidant efficacies on DPPH (IC_{50} 0.227 $mgmL^{-1}$) and Fe^{2+} chelating (IC_{50} 0.157 $mgmL^{-1}$) in comparison to ascorbic acid (IC_{50} 0.109 $mgmL^{-1}$) and EDTA (IC_{50} 0.091 $mgmL^{-1}$) respectively [67]. The HPLC quantification of plant extracts have shown similar classes of compounds such as chlorogenic acid, ellagic acid, caffeic acid, gallic acid, p-coumaric acid, apigenin, quercetin, rutin and kaempferol which have been repeatedly found in Nigerian plants. But the antioxidant efficacies observed were not reflective of HPLC quantification. This may indicate that phenolic compounds are quantified at miniature level which can only serve as evidence of qualitative presence in plant extracts.

7. Antioxidant activities of isolated compounds

The antioxidant evaluations on isolated compounds from Nigerian medicinal plants are rarely reported. This is probably due to funding problems associated to plant chemistry research in Nigeria, coupled with dysfunctional analytical instruments such as the NMR spectrometer. Most of the published research on isolation and characterization of compounds were carried out abroad. Of the 250 plants analyzed for antioxidant evaluations, only 28 compounds were isolated from 44 plants together with full spectral characterization. The antioxidant activities of quercetin and quercetin-3-*O*-rutinoside from *B. monandra* were probably the first report on pure compounds [93]. Since then several isolated compounds were evaluated for antioxidant efficacies and in most cases compared with standard antioxidants. Thus, compounds' efficacy only with IC_{50} values of standards are presented in **Table 2**. The analysis of isolated compounds showed that flavonoids and

S. No	Chemical name	Plant	Model	Compd. (IC_{50})	Stand. (IC_{50})	Ref.
1	Quercetin	*Bauhinia monandra*	DPPH	$10.64^{\wedge}$	AA = 12.52	[93]
2	Quercetin-3-O-rutinoside	*Bauhinia monandra*	DPPH	$16.11^{\wedge}$	AA = 12.52	[93]
3	Isovitexin	*Croton zambesicus*	DPPH	$189.1^{\wedge}$	QT = 5.31	[98]
4	Trans-ethyl-3-(3, 4-dihydroxyphenyl acrylate	*Aspilia africana*	DPPH	$14.49^{\wedge}$	AA = 13.18	[99]
5	p-hydroxy benzaldehyde	*Aspilia africana*	DPPH	73.50^{*}	VC = 37.5	[102]
6	Tiliroside	*Croton gratissimus*	DPPH	360.1^{*}	AA = 70.12	[100]
7	Isovitexin	*Croton gratissimus*	DPPH	211.6^{*}	AA = 70.12	[100]
8	Helichrysoside-3′-methyl ether	*Croton zambesicus*	DPPH	183.4^{*}	AA = 70.12	[100]
9	Betulin	*Parinari curatellifolia*	DPPH	$>100^{*}$	VC = 1.98	[88]
10	β-sitosterol	*Parinari curatellifolia*	DPPH	$>50^{*}$	VC = 1.98	[88]
11	Betulinic acid	*Parinari curatellifolia*	DPPH	$>100^{*}$	VC = 1.98	[88]
12	4-(3′,3-dihydroxyl-1-mercaptopropyl) phenyl-glucosylpyranoside	*Massularia acuminata*	DPPH	75^{*}	VC = 7.59	[87]
13	Agathisflavone	*Anacardium occidentale*	DPPH	366.4^{*}	AA = 4.57	[74]
14	Quercetin-3-O-rutinoside/ rhamnoside	*Anacardium occidentale*	DPPH	0.96^{*}	AA = 4.57	[74]
15	Rosmarinic acid	*Solenostemon monostachyus*	DPPH	4.99^{*}	QT = 2.32	[97]
16	Methyl rosmarinate	*Solenostemon monostachyus*	DPPH	5.97^{*}	QT = 2.32	[97]
17	Caffeic acid	*Solenostemon monostachyus*	DPPH	3.03^{*}	QT = 2.32	[97]
18	Methyl caffeate	*Solenostemon monostachyus*	DPPH	13.41^{*}	QT = 2.32	[97]
19	Apigenin	*Solenostemon monostachyus*	DPPH	26.67^{*}	QT = 2.32	[97]
20	Luteolin	*Solenostemon monostachyus*	DPPH	5.35^{*}	QT = 2.32	[97]
21	Apigenin glucuronide	*Solenostemon monostachyus*	DPPH	185.89^{*}	QT = 2.32	[97]
22	Epicatechin	*Chrysophyllum albidum*	DPPH	$19.02^{\wedge}$	GA = 12.82	[101]

S. No	Chemical name	Plant	Model	Compd. (IC_{50})	Stand. (IC_{50})	Ref.
23	Epigallocatechin	*Chrysophyllum albidum*	DPPH	15.88^	GA = 12.82	[101]
24	Procyanidin B5	*Chrysophyllum albidum*	DPPH	8.80^	GA = 12.82	[101]
25	Kaempferol-3-O-rutinoside	*Holarrhena floribunda*	FRAP	394.8*	QT = 2.95	[85]
26	Quercetin-3-O-glucoside	*Holarrhena floribunda*	LPI FRAP	10.4* 1649.4*	QT = 2.95	[85]
27	Kaempferol-3-O-glucoside	*Holarrhena floribunda*	FRAP	337.5*	QT = 2.95	[85]
28	Quercetin-3-O-glucoside/ galactoside mixture (1: 1)	*Holarrhena floribunda*	LPI FRAP	9.8* 1589.9*	QT = 2.95	[85]
29	Quercetin	*Cassia sieberiana*	DPPH ABTS	1.58# 0.81#	AA = 2.44 TX = 0.81	[96]
30	Kaempferol	*Cassia sieberiana*	DPPH	7.75#	AA = 2.44	[96]
31	Dihydrokaempferol	*Cassia sieberiana*	DPPH	82.93#	AA = 2.44	[96]
32	Piceatannol	*Cassia sieberiana*	DPPH	3.96#	AA = 2.44	[96]
33	(−)-Catechin	*Alchornea floribunda*	DPPH H_2O_2	88* 13*	AA = 6 AA = 8	[95]
34	(+)-epicatechin	*Alchornea floribunda*	DPPH H_2O_2	40* 10*	AA = 6 AA = 8	[95]
35	(−)-epicatechin	*Alchornea floribunda*	DPPH H_2O_2	10* 8*	AA = 6 AA = 8	[95]
36	(2R,3R)-dihydroquercetin	*Alchornea floribunda*	DPPH H_2O_2	46* 18*	AA = 6 AA = 8	[95]
37	Catechin	*Annona senegalensis*	DPPH Fe(II)	0.03** 1.29**	AA = 0.01 EDTA = 0.05	[15]

DPPH, 2,2-diphenyl-1-picrylhydrazyl radical; ABTS, 2,2'-azinobis (3-ethylbenzothiazoline-6-sulfonic acid; FRAP, ferric reducing antioxidant power; TAC, total antioxidant capacity; LPI, lipid peroxidation inhibition; NO, nitric oxide assay; H_2O_2, hydrogen peroxide assay; AA, ascorbic acid; QT, quercetin; RT, rutin; GA, gallic acid; VC, vitamin C; VE, vitamin E; TX, trolox; EDTA: ethylenediaminetetraacetic acid.
^IC_{50} = μM.
#IC_{50} = mM.
**IC_{50} = μgmL^{-1}.*
***IC_{50} = mgmL^{-1}.*

Table 2.
Antioxidant activity of isolated compounds of Nigerian plants.

flavonoids glycosides constitute major classes of antioxidants reported. Catechin isolated from *A. senegalensis* had effective DPPH inhibition (IC_{50} 0.03 mgmL^{-1}) and Fe^{2+} chelating activity (1.29 mgmL^{-1}) when compared to ascorbic acid (0.01 mgmL^{-1}) and EDTA (IC_{50} 0.05 mgmL^{-1}) respectively [94]. The evaluation on H_2O_2 inhibition by (−)-epicatechin isolated from *A. floribunda* showed effective activity with equal strength as standard ascorbic acid (IC_{50} 8.0 μgmL^{-1}) [95]. Similarly, the ABTS inhibition by quercetin isolated from *C. sieberiana* has resulted to effective activity of equal strength to Tocopherol (0.81 mM) [96]. The DPPH inhibition by caffeic acid (IC_{50} 3.03 μgmL^{-1}) from *S. monostachys* is another effective activity comparable to quercetin standard (IC_{50} 2.32 μgmL^{-1}) [97]. However, the most outstanding DPPH inhibition was recorded on quercetin-3-*O*-rutinoside/rhamnoside isolated from *A. occidentalis*. The 1:1 mixture of flavonoid glycoside exhibited

IC_{50} 0.96 μgmL^{-1} less than ascorbic acid (IC_{50} 4.57 μgmL^{-1}). [74]. But generally, the antioxidant efficacies of isolated compounds from Nigerian plants are not interesting. Out of the 28 compounds isolated from 44 plants only 7 compounds from 6 plants exhibited the efficacies with strength of standard antioxidants.

8. Conclusion

Analysis of antioxidant efficacies of Nigerian medicinal plants reported from 1998 to 2018 was carried out. The aim was to provide evidence for effective antioxidants. Our findings have shown the enormous potentials of Nigerian plants as sources of natural antioxidants. We have observed various crude extracts obtained mainly from polar solvents with antioxidant efficacies better than standard compounds. Such preponderance of evidence indicated by broad spectrum of free radical and non-free radical inhibitions has defined the comparable strength of plant extracts to standard antioxidants. Nigerian plants have the capacity to protect or inhibit damage induced by free radical species. This study attempts to provide insights on the strength of antioxidant efficacies of plant extracts comparable to standard antioxidants. However, it is recommended that comprehensive approach to plant bioactive research must be adopted in search of antioxidants to avoid replication of studies especially on certain species. There is need for collaboration among Nigerian scientist working in related areas to enhance on the scope of research questions and improve on the quality of research output.

Acknowledgements

Authors acknowledged the plant taxonomist, Umar Shehu Gallah of National Research Institute for Chemical Technology (NARICT) Zaria, for providing the local (Hausa) names of plants. We are grateful to Ahmadu Bello University, Zaria for providing some of the facilities used during the study.

Author details

Abubakar Babando Aliyu*, Jonathan Ilemona Achika, Joseph Adesina Adewuyi, Patience Gangas, Hamisu Ibrahim and Adebayo Ojo Oyewale
Organic Chemistry Research Laboratory, Department of Chemistry, Faculty of Physical Sciences, Ahmadu Bello University, Zaria, Nigeria

*Address all correspondence to: aliyubabando@gmail.com

References

[1] McCord JM, Fridovich I. Superoxide dismutase an enzymic function for erythrocuprein Hemocuprein. Journal of Biological Chemistry. 1969;**244**:6049-6055

[2] Prior RL, Wu X, Schaich K. Standardized methods for the determination of antioxidant capacity and phenolics in foods and dietary supplements. Journal of Agriculture and Food Chemistry. 2005;**53**:4290-4302

[3] Carocho M, Ferreira ICFR. A review on antioxidants, prooxidants and related controversy: Natural and synthetic compounds, screening and analysis methodologies and future perspectives. Food and Chemical Toxicology. 2013;**51**:15-25

[4] West AP, Shadel GS, Ghosh S. Mitochondria in innate immune responses. Nature Reviews Immunology. 2011;**11**:389-402

[5] Kurutas EB. The importance of antioxidants which play the role in cellular response against oxidative/nitrosative stress: Current state. Nutrition Journal. 2016;**15**:71

[6] Sanchez C. Reactive oxygen species and antioxidant properties from mushrooms. Synthetic and Systems Biotechnology. 2016;**2**:13e22

[7] Halliwell B. Free radicals, antioxidants, and human disease: Curiosity, cause, or consequence? Lancet. 1994;**344**:721e4

[8] Kozarski M, Klaus A, Jakovljevic D, Todorovic N, Vunduk J, Petrovic P. Antioxidants of edible mushrooms. Molecules. 2015;**20**:19489e525

[9] Cederbaum AI, Lu Y, Wu D. Role of oxidative stress in alcohol-induced liver injury. Archives of Toxicology. 2009;**83**(6):519e48

[10] George S, Hamblin MR, Abrahamse H. Effect of red light and near infrared laser on the generation of reactive oxygen species in primary dermal fibroblasts. Journal of Photochemistry & Photobiology, B: Biology. 2018;**188**:60-68

[11] Qian X, Nie X, Yao W, Klinghammer K, Sudhoff H, Kaufmann AM, et al. Reactive oxygen species in cancer stem cells of head and neck squamous Cancer. Seminars in Cancer Biology. 2018. DOI: 10.1016/j.semcancer.2018.06.001

[12] Sies H. Strategies of antioxidant defense. FEBS Journal. 1993;**215**:213-219

[13] Halliwell B. Biochemistry of oxidative stress. Biochemical Society Transaction. 2007;**35**:1147-1150

[14] Halliwell B. How to characterize a biological antioxidant. Free Radical Research Communication. 1990;**9**:1-32

[15] Govindarajan R, Vijayakumar M, Pushpangadan P. Antioxidant approach to disease management and the role of 'Rasayana' herbs of Ayurveda. Journal of Ethnopharmacology. 2005;**99**:165-178

[16] Larson RA. The antioxidants of higher plants. Phytochemistry. 1988;**27**(4):969-978

[17] Kahkonen MP, Hopia AI, Vuorela HJ, Rauha JP, Pihlaja K, Kujala TS, et al. Antioxidant activity of plant extracts containing phenolic compounds. Journal of Agriculture and Food Chemistry. 1999;**47**(10):3954-3962

[18] Nakayoma J, Yamada M. Suppression of active oxygen-indeed cytotoxicity by flavonoids. Biochemical Pharmacology. 1995;**45**:265-267

[19] Bhadoriya U, Sharma P, Solanki SS. *In vitro* free radical scavenging activity of gallic acid isolated from *Caesalpinia decapetala* wood. Asian Pacific Journal of Tropical Disease. 2012:S833-S836

[20] Xu DP, Li Y, Meng X, Zhou T, Zhou Y, Zheng J, et al. Natural antioxidants in foods and medicinal plants: Extraction, assessment and resources. International Journal of Molecular Science. 2017;**18**:96

[21] National Bureau of Statistics. Annual Abstract of Statistics. Abuja, Nigeria: Federal Republic of Nigeria; 2011. p. 3

[22] Aderogba MA, Okoh EK, Adelawa TA, Obuotor EM. Antioxidant properties of the Nigerian *Piliostigma* species. Journal of Biological Sciences. 2004;**4**(4):501-503

[23] Ibrahim T, Muluh GK, Alexander A. Phytochemical screening, antioxidant and antibacterial activities of *Commiphora kerstingii*. International Biological and Biomedical Journal. 2015;**2**(3):126-133

[24] Ogunlana OE, Ogunlana OO. *In vitro* assessment of the free radical scavenging activity of *Psidium guajava*. Research Journal of Agriculture and Biological Sciences. 2008;**4**(6):666-671

[25] Azeez L, Adeoye MD, Majolagbe TA, Lawal AT, Badiru R. Antioxidant activity and phytochemical contents of some selected Nigerian fruits and vegetables. American Journal of Chemistry. 2012;**2**(4):209-213

[26] Adetutu A, Morgana WA, Corcoran O. Ethnopharmacological survey and in vitro evaluation of wound-healing plants used in South-Western Nigeria. Journal of Ethnopharmacology. 2011;**137**:50-56

[27] Olajire AA, Azeez L. Total antioxidant activity, phenolic, flavonoid and ascorbic acid contents of Nigerian vegetables. African Journal of Food Science and Technology. 2011;**2**(2):22-29

[28] Atiko R, Onocha PA, Oyedemi SO. Phytochemical analysis and antioxidant properties of leaves, stems and roots extracts of *Lasianthera africana*. Research Journal of Chemical Sciences. 2016;**6**(9):19-26

[29] Ekaluo UB, Ikpeme EV, Ekerette EE, Chukwu CI. *In vitro* antioxidant and free radical activity of some Nigerian medicinal plants: Bitter leaf (*Vernonia amygdalina* L.) and guava (*Psidium guajava* Del.). Research Journal of Medicinal Plant. 2015;**9**(5):215-226

[30] Sonibare MA, Aremu OT, Okorie PN. Antioxidant and antimicrobial activities of solvent fractions of *Vernonia cinerea* (L.) leaf extract. African Health Sciences. 2016;**16**(2):629-639

[31] Abalaka ME, Daniyan SY, Akpor OB, Inyinbor AA. Antimicrobial, *in-vitro* free radical scavenging, antioxidant properties of leaf, bark and root extracts from *Khaya senegalensis*. UMYU Journal of Microbiology Research. 2016;**1**(1):46-54

[32] Adebiyi OE, Olayemi FO, Ning-Hua T, Guang-Zhi Z. *In vitro* antioxidant activity, total phenolic and flavonoid contents of ethanol extract of stem and leaf of *Grewia carpinifolia*. Beni-Suef University Journal of Basic and Applied Sciences. 2017;**6**:10-14

[33] Adeosun AM, Ighodaro OM, Aminu AO, Ogunlana AI. The antioxidant and phenolic profiles of five green vegetables grown in southern Nigeria. Acta Scientiarum Polonorum Technologia Alimentaria. 2016;**15**(4):391-397

[34] Dluya T, Daniel D, Umar Y. *In vitro* antioxidant activity and phytochemical evaluation of five medicinal plant

extracts. The Pharmaceutical and Chemical Journal. 2017;**4**(5):73-82

[35] Dluya T, Daniel D, Samuel AB. Studies on *in vitro* antioxidant activities of methanol extract of five African traditional plants. International Journal of Research in Pharmacy and Biosciences. 2017;**4**(11):28-36

[36] Antia BS, Ita BN, Udo UE. Nutrient composition and *in vitro* antioxidant properties of *Harungana madagascariensis* stembark extracts. Journal of Medicinal Food. 2015;**18**(5):609-614

[37] Okoh SO, Asekun OT, Familoni OB, Afolayan AJ. Antioxidant and free radical scavenging capacity of seed and shell essential oils extracted from *Abrus precatorius* (L). Antioxidants. 2014;**3**:278-287

[38] Nwaehwujor OC, Ode JM, Akande OG. *In vitro* antioxidant activity of some herbal plants from Southern Nigeria. Journal of Medical Science. 2013;**1**(13):56-611

[39] Sidiq LO, Segun PA, Ogbole OO. Total phenolic contents and antioxidant activity of nine medicinal plants used in nigerian traditional medicine. Tropical Journal of Natural Product Research. 2018;**2**(9):438-441

[40] Ekere N, Okparanozie T, Agbo M. Effects of solvents on the *in-vitro* antioxidant activity of *Dennettia tripetala* G. Baker and *Milicia excelsa* (Welw.) C. Berg root extracts. Iranian Journal of Health Sciences. 2018;**6**(3):1-7

[41] Ohikhena FU, Wintola OA, Afolayan AJ. Quantitative phytochemical constituents and antioxidant activities of the mistletoe, *Phragmanthera capitata* (Sprengel) Balle extracted with different solvents. Pharmacognosy Research. 2018;**10**:16-23

[42] Osagie-Eweka SDE. Phytochemical analyses and comparative *in vitro* antioxidant studies of aqueous, methanol and ethanol stem bark extracts of *Simarouba glauca* DC. (Paradise tree). African Journal of Plant Science. 2018;**12**(1):7-16

[43] Owolabi OO, James DB, Sani I, Andongma BT, Fasanya OO, Kure B. Phytochemical analysis, antioxidant and anti-inflammatory potential of *Feretia apodanthera* root bark extracts. BMC Complementary and Alternative Medicine. 2018;**18**:12. DOI: 10.1186/s12906-017-2070-z

[44] Muhamad II, Hassan ND, Mamat SNH, Nawi NM, Rashid WA, Tan NA. Chapter 14: Extraction technologies and solvents of phytocompounds from plant materials: Physicochemical characterization and identification of ingredients and bioactive compounds from plant extract using various instrumentations. In: Ingredients Extraction by Physicochemical Methods in Food. Elsevier Inc.; 2017. pp. 523-560. DOI: 10.1016/B978-0-12-811521-3.00014-4

[45] Ajiboye AT, Musa MD, Otun KO, Jimoh AA, Bale AT, Lawal SO, et al. The studies of antioxidant and antimicrobial potentials of the leaf extract of *Bauhinia monandra* plant. Natural Products Chemistry and Research. 2015;**3**:180. DOI: 10.4172/2329-6836.1000180

[46] Akachukwu D, Uchegbu R. GC-MS antimicrobial and *in vitro* antioxidant assay of the leaf extract of. Journal of Advances in Medical and Pharmaceutical Sciences. 2016;**11**(2):1-7

[47] Mohammed A, Koorbanally NA, Islam S. Phytochemistry, antioxidative and antidiabetic effects of various parts of *Eugenia caryophyllata* thunb. *in vitro*. Acta Poloniae Pharmaceutica-Drug Research. 2015;**72**(6):1201-1215

[48] Ibrahim MA, Koorbanally NA, Shahidul Islam MD. *In vitro* anti-oxidative activities and GC-MS analysis of various solvent extracts of *Cassia singueana* parts. Acta Poloniae Pharmaceutica-Drug Research. 2013;**70**(4):709-719

[49] Iwara IA, Igile GO, Mboso OE, Mgbeje BIA, Ebong PE. Evaluation of phytochemical components from ethyl acetate fraction of *Vernonia calvoana* using gas chromatography-mass spectrometry analysis and its antioxidants activities. African Journal of Pharmacy and Pharmacology. 2017;**11**(42):534-539

[50] Atolani O, Olatunji GA. Chemical composition, antioxidant and cytotoxicity potential of *Daniellia oliveri* (Rolfe) Hutch. & Dalz. Turkish Journal of Pharmaceutical Sciences. 2016;**13**(1):41-46

[51] Ololade ZS, Ogunmola OO, Kuyooro SE, Abiona OO. *Stachytarpheta jamaicensis* leaf extract: Chemical composition, antioxidant, anti-arthritic, anti-inflammatory and bactericidal potentials. Journal of Scientific and Innovative Research. 2017;**6**(4):119-125

[52] Elufioye TO, Berida TI. GC-MS analysis and antioxidant activity of *Spondias purpurea* L (Anacardiaceae). Pharmacognosy Journal. 2018;**10**(5): 941-945

[53] Ogundajo A, Ashafa AT. Phytochemical compositions and *In vitro* assessments of antioxidant and antidiabetic potentials of fractions from *Ehretia cymosa* Thonn. Pharmacognosy Magazine. 2017;**13**(S3):470-480

[54] Fasola TR, Oloyede GK, Aponjolosun SB. Chemical composition, toxicity and antioxidant activities of essential oils of stem bark of Nigerian species of guava (*Psidium guajava* Linn.). EXCLI Journal. 2011;**10**:34-43

[55] Ololade ZS, Fakankun OA, Alao FO, Udi OU. *Ocimum basilicum* var. *purpureum* floral essential oil: phytochemicals, phenolic content, antioxidant, free radical scavenging and antimicrobial potentials. Global Journal of Science Frontier Research: B Chemistry. 2014;**14**(7):31-38

[56] Owokotomo IA, Jabar JM, Abata EO. Radical scavenging activity of essential oils from some Nigerian medicinal plants and spices. Applied Tropical Agriculture. 2016;**21**(1):196-200

[57] Okhale SE, Okoro IJ, Ezissi CC, Oladosu PO. GC-MS characterization, antimicrobial and antioxidant effects of the leaf essential oil of *Calliandra portoricensis* (Jacq.) Benth. International Journal of Pharmaceutical Science and Research. 2018;**3**(1):38-43

[58] Onocha A, Oloyede GK, Akintola JA. Chemical composition, free radical scavenging and antimicrobial activities of essential oil of *Mariscus alternifolius* Vahl Patricia. The Open Conference Proceedings Journal. 2016;7:160-167

[59] Ololade ZS, Olawore NO, Oladosu IA. Phytochemistry and therapeutic potentials of the seed essential oil of *Eucalyptus maculata* Hook from Nigeria. Organic Chemistry Current Research. 2013;**2**:1

[60] Boligon AA, Athayde ML. Importance of HPLC in analysis of plants extracts. Austin Chromatography. 2014;**1**(3):1-2

[61] Adekunle AS, Kamdem JP, Rocha JBT. Antioxidant activity and HPLC analysis of *Zanthozylum zanthoxyloide*. Report and Opinion. 2012;**4**(3):6-13

[62] Borokini FB, Labunmi L. *In vitro* investigation of antioxidant activities of *Launea taraxacifolia* and *Crassocephalum rubens*. International Journal of Food Studies. 2017;**6**:82-94

[63] Piluzza G, Bullitta S. Correlations between phenolic content and antioxidant properties in twenty-four plant species of traditional ethnoveterinary use in the Mediterranean area. Pharmaceutical Biology. 2011;**49**(3):240-247

[64] Kamdem JP, Stefanello ST, Boligon AA, Wagner C, Kade IJ, Pereira RP, et al. *In vitro* antioxidant activity of stem bark of *Trichilia catigua* Adr. Juss. Acta Pharmacetica. 2012;**62**:371-382

[65] Obafemi TO, Akinmoladun AC, Olaleye MT, Onasanya A, Komolafe KC, Falode JA, et al. High performance liquid chromatography (HPLC) fingerprinting, mineral composition and *in vitro* antioxidant activity of methanol leaf extract of *Synsepalum dulcificum* (sapotaceae). Journal of Applied Pharmaceutical Science. 2017;**7**(11):110-118

[66] Esievo KB, Fatokun OT, Adamu A, Egharevba HO. HPLC analysis, anti-oxidant and antiproliferative evaluation of methanol extracts of leaves and roots of *Mondia whitei* (Hook. f) Skeels. Journal of Chemical and Pharmaceutical Research. 2018;**10**(4):81-87

[67] Kadiri O, Akanbi CT, Olawoye BT, Saka O, Gbadamosi SO. Characterization and antioxidant evaluation of phenolic compounds extracted from the protein concentrate and protein isolate produced from pawpaw (*Carica papaya* Linn.) seeds. International Journal of Food Properties. 2017;**20**(11):2423-2436

[68] Onocha PA, Oloyede GK, Olasunkanmi GS. Chemical composition, brine shrimp toxicity and free-radical scavenging activity of leaf essential oil of *Acalypha Ornata* (Hochst). Advances in Environmental Biology. 2011;**5**(1):188-193

[69] Mohammed A. Antioxidative and antidiabetic effects of some African medicinal plants [PhD Biochemistry]. School of Life Sciences, College of Agriculture, Engineering and Science, University of KwaZulu-Natal, Durban, South Africa. 2016. pp. 1-250

[70] Omonhinmin AC, Dike IP, Rotimi E. Phytochemical, cytotoxicity and antioxidant activities of five anti-malaria plants. Research Journal of Medicinal Plant. 2015. DOI: 10.3923/rjmp.2015.81.89

[71] Ayoola GA, Coker HAB, Adesegun SA, Adepoju-Bello AA, Obaweya K, Ezennia EC, et al. Phytochemical screening and antioxidant activities of some selected medicinal plants used for malaria therapy in Southwestern Nigeria. Tropical Journal of Pharmaceutical Research. 2008;**7**(3):1019-1024

[72] Omoregie ES, Oriakhi K, Oikeh EI, Okugbo OT, Akpobire A. Comparative study of phenolic content and antioxidant activity of leaf extracts of *Alstonia boonei* and *Eupatorium odoratum*. Nigerian Journal of Basic and Applied Science. 2014;**22**(3&4):91-97

[73] Awah FM, Uzoegwu PN, Ifeonu P, Oyugi JO, Rutherford J, Yao XJ, et al. Free radical scavenging activity, phenolic contents and cytotoxicity of selected Nigerian medicinal plants. Food Chemistry. 2012;**131**:1279-1286

[74] Ajileye OO, Obuotor EM, Akinkunmi EO, Aderogba MA. Isolation and characterization of antioxidant and antimicrobial compounds from *Anacardium occidentale* L. (Anacardiaceae) leaf extract. Journal of King Saud University-Science. 2015;**27**:244-252

[75] Uche FI. Evaluation of antioxidant activities of some Nigerian medicinal plants by the use of 2,2-diphenyl-1-picrylhydrazyl (DPPH) free radical. MSc Pharmaceutical Sciences Degree in Pharmacognosy. Department

of Pharmacognosy Faculty of Pharmaceutical Sciences, University of Nigeria, Nsukka. 2010. p. 80

[76] Aderogba MA, McGraw LJ, Ogundaini AO, Eloff JN. Antioxidant activity and cytotoxicity study of the flavonol glycosides from *Bauhinia galpinii*. Natural Product Research. 2007;**21**(7):591-599

[77] Adesegun SA, Orabueze CI, Coker HAB. Antimalarial and antioxidant potentials of extract and fractions of aerial part of *Borreria ocymoides* DC (Rubiaceae). Pharmacognosy Journal. 2017;**9**(4):534-540

[78] Gero A, Ahmad HS, Zezi AU, Hussaini IM. Evaluation of antioxidant activity of leave extract of *Borreria verticillta* Linn (Rubiaceae). Journal of Natural Sciences Research. 2014;**4**(9):31-38

[79] Faboro EO, Witchitnihad W, Fadere OA, Akinpelu DA, Obefemi CA. Antibcterial and antioxidant and phytochemical screening of aqueous methanol extract of eight Nigerian medicinal and aromatic plants. Journal of Pharmacy Research. 2016;**10**(7):523-532

[80] Falodun A, Irabor EEI. Phytochemical, proximate, antioxidant and free radical scavenging evaluations of *Calliandria surinamensis*. Acta Poloniae Pharmaceutica-Drug Research. 2008;**65**(5):571-575

[81] Famobuwa OE, Adekunbi EA, Akinnifesi TA, Hassan GF. *In vitro* antioxidant and anti-bacterial properties of a contraceptive herbal mixture of *Zanthoxylum zanthoxyloides* Lam (Rutaceae), *Euphorbia hirta* Linn (Euphorbiaceae) and *Abrus precatorius* L. (Leguminosae). Journal of Advances in Medical and Pharmaceutical Sciences. 2016;**7**(1):1-7

[82] Atiku I, Pateh UU, Musa AM, Sule MI, Sani YM, Abdullahi SM, et al. Phytochemical and antioxidant activity studies of the ethanol leaf extract of *Ficus sycomorus* (Family: Moraceae). Nigerian Journal of Pharmaceutical Sciences. 2016;**15**(2):8-14

[83] Essien GE, Effiong GS. Evaluation of antioxidant potentials in methanol extract of *Gongronema latifolium* and *Lasiathera africana* leaf. European Journal of Pharmaceutical and Medical Research. 2017;**4**(9):841-846

[84] Kambizi L, Bakare-Odunola MT, Oladiji AT, Kola-Mustapha AT, Amusa TO, Atolani O, et al. Proteinase inhibition, membrane stabilization, antioxidant and phytochemical evaluations of leaves, seeds and calyces of four selected edible medicinal plants. Cogent Chemistry. 2017;**3**:1314064

[85] Badmus JA, Ekpo OE, Rautenbach F, Marnawick JL, Hussein AA, Hiss DC. Isolation and antioxidant activity of flavonoids from *Holarrhena floribunda* (G.don) leaves. Acta Biochemica Polonica. 2016;**63**(2):353-358

[86] Onoja OS, Ezeja MI, Omeh YN, Onwukwe BC. Antioxidant, anti-inflammatory and antinociceptive activities of methanolic extract of *Justicia secunda* Vahl leaf. Alexandria Journal of Medicine. 2017;**53**:207-213

[87] Oriola AO, Aladesanmi AJ, Idowu TO, Akinkunmi EO, Obuotor EM, Ogunsina MO. A new bioactive thiophenolic glycoside from the leaf of *Massularia acuminata* (g. don bullock) ex hoyle (Rubiaceae). African Journal of Traditional, Complementary and Alternative Medicine. 2014;**11**(2):319-323

[88] Halilu ME, October N, Balogun M, Namrita L, Abubakar MS. Studies of *in vitro* antioxidant and cytotoxic activities of extracts and isolated

compounds from *Parinari curatellifolia* (Chrysobalanaceae). Journal of Natural Sciences Research. 2013;**3**(13):149-154

[89] Adesegun SA, Elechi NA, Coker HAB. Antioxidant activities of methanol extract of *Sapium ellipticum*. Pakistan Journal Biological Sciences. 2008;**11**(3):453-457

[90] Ojo OA, Ojo AB, Ajiboye BO, Olaiya O, Okesola MA, Boligon AA, et al. HPLC-DAD fingerprinting analysis, antioxidant activities of *Tithonia diversifolia* (Hemsl.) A. Gray leaves and its inhibition of key enzymes linked to Alzheimer's disease. Toxicology Reports. 2018;**5**:585-592

[91] Halilu ME, Usman AO, Sani NA, Hassan LG, Ugwuah-Oguejiofor CJ, Bello SS. Evaluation of free radical scavenging activity of *Vernonia migeodii* (S. Moore) leave extracts. Nigerian Journal of Pharmaceutical and Biomedical Research. 2016;**1**(1):8-13

[92] Yusuf AA, Lawal B, Abubakar AN, Berinyuy EB, Omonije YO, Umar SI, et al. *In-vitro* antioxidants, antimicrobial and toxicological evaluation of Nigerian *Zingiber officinale*. Clinical Phytoscience. 2018;**4**:12

[93] Aderogba MA, Ogundaini AO, Eloff JN. Isolation of two flavonoids from *Bauhinia monandra* (kurz) leaves and their antioxidative effects. African Journal of Traditional, Complementary and Alternative Medicine. 2006;**3**(4):59-65

[94] Aladejare TT, Obuotor EM, Aderogba MA. Isolation and characterization of antioxidant compounds from *Annona senegalensis* leaf extracts. Nigerian Journal of Natural Products and Medicine. 2018;**22**:124-128

[95] Ajaghaku DL, Obasi O, Umeokoli BO, Ogbuatu P, Nworu CS, Ilodigwe EE, et al. *In vitro* and *in vivo* antioxidant potentials of *Alchornea floribunda* leaf extract, fractions and isolated bioactive compounds. Avicenna Journal of Phytomedicine. 2017;**7**(1):80-92

[96] Jibril S, Sirat HM, Basar N. Bioassay-guided isolation of antioxidants and α-glucosidase inhibitors from the root of *Cassia sieberiana* D.C. (Fabaceae). Records of Natural Product. 2017;**11**(4):406-410

[97] Taiwo BJ, Obuotor EM, Onawunmi GO, Oguntuga AO. Radical scavenging compounds from the aerial parts of *Solenostemon monostachys* BRIQ (LAMIACEAE). African Journal of Traditional Complementary and Alternative Medicine. 2015;**12**(6):140-144

[98] Aderogba MA, McGawc LJ, Bezabihb M, Abegazb BM. Isolation and characterization of novel antioxidant constituent of *Croton zambesicus* leaf extract. Natural Product Research. 2011;**25**(13):1224-1233

[99] Jide FF, Abiodun OO. Evaluation of antioxidant and antimicrobial properties of two isolates from *Aspilia africana*. International Research Journal of Pharmacy. 2012;**3**(7):135-138

[100] Ndhlala AR, Aderogba MA, Ncube B, van Staden J. Antioxidative and cholinesterase inhibitory effects of leaf extracts and their isolated compounds from two closely related *Croton* species. Molecules. 2013;**18**:1916-1932

[101] Idowu TO, Ogundaini AO, Adesanya SA, Onawunmi GO, Osungunna MO, Obuotor EM, et al. Isolation and characterization of chemical constituents from *Chrysophyllum albidum* stem-bark extracts and their antioxidant and antibacterial properties. African Journal of Traditional Complementary and

Alternative Medicine. 2016;**13**(5): 182-189

[102] Johnson EC, Etim EI, Archibong EO. Isolation and anti-oxidant potentials of para-hydroxybenzaldehyde from the methanol leaf extract of *Aspilia africana* (Pers.) C.D. Adams (Asteraceae). Nigerian Journal of Pharmaceutical and Applied Science Research. 2017;**6**(1):26-32

Chapter 5

Royal Jelly and Human Interferon-Alpha (HuIFN-αN3) Affect Proliferation, Glutathione Level, and Lipid Peroxidation in Human Colorectal Adenocarcinoma Cells In Vitro

Bratko Filipič, Lidija Gradišnik, Klemen Rihar, Adriana Pereyra, Damir Đermić and Hrvoje Mazija

Abstract

The purpose was to investigate the influence of RJ-F(M), 10-hydroxy-2-decenoic acid and HuIFN-αN3 on the proliferation of CaCo-2 cells and ascertain their effects on intracellular glutathione level and lipid peroxidation. The antiproliferative (AP) activity of RJ-F (M) (0.1 g/10 mL PBS), HuIFN-αN3 (1000 IU mL^{-1}), 10-HDA (100.0 μmol L^{-1}) and their combinations, in the ratios 1:1, 1:2, and 2:1 on CaCo-2 cells were measured. Single RJ-F (M) had a low AP activity: 2.0 (0.5 mg mL^{-1}). HuIFN-αN3 had an AP activity of 2.5 (208.33 IU mL^{-1}), while 10-HDA had an AP activity of 1.5 (37.5 μmol mL^{-1}). AP activity of 3.8 was obtained when RJ-F(M) and HuIFN-αN3 were in the ratio 2:1. In it, the level of GSH was 24.9 ± 2.4 nmol g^{-3} of proteins (vs. 70.2 ± 3.2 nmol g^{-3} in the control), and level of MDA was 72.3 ± 3.1 nmol g^{-3} (vs. 23.6 ± 9.1 nmol g^{-3} in the control). 10-HDA, together with HuIFN-αN3, is responsible for the inhibition of CaCo-2 cell proliferation *in vitro*. RJ-F (M) and HuIFN-αN3 applied at 2:1 decreased level of GSH and increased lipid peroxidation *via* MDA in CaCo-2 cells. Future studies are needed whether these GSH- and MDA-related activities of RJ-F (M), HuIFN-αN3, 10-HDA, and their combinations may decrease the tumorigenicity index and tumorigenic potential of tumor cells *in vitro*

Keywords: antiproliferative activity, antitumor activity, CaCo-2 cells, 10-hydorxy-2-decenoic acid, malondialdehyde, HuIFN-αN3, RJ-F (M)

1. Introduction

Royal jelly (RJ) is a milky acid colloid of a pH between 3.6 and 4.2, produced by hypo pharyngeal and mandible glands of young worker bees between sixth and twelfth days of their life. Fed as a sole nutrient, it is the exclusive food for the bee colonies' queen honey bee (*Apis mellifera*) larva, developing into a sexual mature queen bee. RJ has a crucial role in the life span of bees: worker bee lives around 45 days, while a queen bee can be alive up to 5 years. During this time, she is able to spawn daily an equivalent of her body weight in eggs (approximately 2000–3000 eggs/day for up to 5 years) [1, 2]. Chemically, RJ is composed by water (50–60%), different proteins (18.0%), carbohydrates (15.0%), lipids (3–6%), mineral salts (1.5%), and vitamins, together with bioactive substances, such as 10-hydroxy-2-decenoic (10-HDA) acid [3]. 10-HDA exerts an inhibitory effect on vascular endothelial growth factor-induced angiogenesis in cellular models of tumor growth, partly by inhibiting cell proliferation and migration [4]. RJ contains different immunomodulative and antiproliferative substances as well as substances with possible antitumor activities. It also contains proteins, like 350-kDa protein, that stimulate proliferation of human monocytes [5–7]. In addition, RJ $protein_{30}$ can be found; it has shown an prominent cytotoxic effect on HeLa cells in vitro [8]. A protein component of RJ belong to the family of major royal jelly proteins (MRJPs) or an albumin. In this families, eight proteins (MRJP 1–8) were identified and found that they have molecular masses between 49 and 87 kDs. MRJPs have an important role in the diet of the queen bee [9]. Other proteins in RJ that are in much lower amount are royalisin, jelleines, aspimin, and royalactin. Researchers showed antibacterial properties of royalisins against Gram-positive bacteria and proposed their use as potential antimicrobial [10]. Royalisins are composed of 51 residues with a net charge of +2; even their origin is unknown. It was supposed that they could originate directly from honeybee. It was also shown that the RJ's protein royalactin is needed for queen bee differentiation. It activates the epidermal growth factor receptor (EGFR) signaling pathway in honeybee larvae, which leads to queen development. Corona et al. [11] found that royalactin induces queen development through shortening her development time and increases longevity of the queen bee by the EGFR signaling pathway.

In RJ, glucose oxidase enzyme (GOx) was also detected. GOx catalyzing the oxidation of glucose to hydrogen peroxide was showing a relatively high antibacterial activity [12].

Interferons (IFNs) are multifunctional glycoproteins/proteins produced and released by host cells in response to the presence of pathogens such as viruses, bacteria, parasites, or tumor cells. There are three classes of IFNs, designated as Types I, II, and III. HuIFN-αN3, a Type I interferon, is a protein composed from 13 subtypes [13, 14] having antiviral, antiproliferative, and antitumor activities [15]. Type I IFN isoforms were found since its discovery. These, coded by a single exon, include IFN-α (composed from 13 subtypes: IFN-α1, IFN-α2, IFN-α4, IFN-α5, IFN-α6, IFN-α7, IFN-α8, IFN-α10, IFN-α13, IFN-α14, IFN-α16, IFN-α17, and IFN-α21), IFN-β, IFN-ε, IFN-κ, and IFN-ω [16]. The latter was used from the beginning for the treatment of different forms of cancers, despite its molecular mechanism of cytoreductive action being far from clear. HuIFN-αN3's antiproliferative activity plays a central role in its chemotherapeutic effect. The research explained its action in apoptosis as a way of antitumor action [17]. In addition, HuIFN-αN3 shows a direct cytotoxic effect on malignant cells and different tumor cell lines *in vitro* [18]. It was also shown that endogenous type I IFN prevents the growth of primary carcinogen-induced and transplantable tumors [19]. One of the earliest described immunoregulatory functions of HuIFN-αN3 is its ability to regulate NK cell functions that is one of the major mechanisms of regulation of tumor growth by the endogenous type I IFN. The role of HuIFN-αN3 in regulating tumor cell proliferation, apoptosis, and autophagy was only recently begun to be investigated in depth [20]. In addition, the ability of oncogenes such as Ras and HPV16 E6E7 for downregulating the IFN-inducing innate receptors, RIG-I and TLR9, is a suggestion for a role of HuIFN-αN3 in modulating the infectiveness of cell-transforming ability of oncogenic viruses [21].

The low-molecular-weight thiol, glutathione (GSH) with a role in the control of the thiol/disulfide redox state in the cells, is important for cellular redox signaling and also for CaCo-2 cells. Intracellular concentration of GSH, cell proliferation, and apoptosis are mutually connected. GSH in a relatively high level enhances the cell proliferation. In contrary, its depletion results in the inhibition of CaCo-2 cell growth and proliferation, because of the increase of apoptosis [22].

Oxidative degradation of lipids or lipid peroxidation (LP) is a process where free radicals "rob" electrons from cellular membrane lipids and induce cell damage by proceeding with the free radical chain reaction mechanism. The LP's end products are reactive aldehydes with a relatively strong carcinogenic potential.

Malondialdehyde (MDA) is one of the important reactive aldehydes. It is a bioactive marker of the LP with a different biological activity similar to reactive oxygen species [23, 24].

The aim of the present study was to explore the effect of RJ-F (M) and 10-HDA on the HuIFN-αN3-induced inhibition of CaCo-2 cell proliferation in vitro. Also important is to ascertain their influence on the level of the GSH in the cells and also on the lipid peroxidation via the MDA activity. Such an experimental approach might serve to explain some of the antiproliferative/antitumor mechanisms in CaCo-2 cells. All this could be also important for the developing of future antiproliferative treatments based on the use of the mentioned bioactive compounds.

2. Materials and methods

2.1 Materials

During the experiments, the following materials were used: (1) HuIFN-αN3 (Institute of Immunology, Zagreb, Croatia) was applied at 1000 IU mL^{-1}. This concentration was found in our previous experiments [25]. (2) Fresh royal jelly (Mižigoj) (RJ-F (M)) (Medex D.o.o., Ljubljana, Slovenia) was applied in a concentration of 0.1 g/10 mL [26]. (3) 10-hydroxy-2-decenoic acid (10-HDA) (Sigma-Aldrich, Missouri, USA) was applied in a concentration of 100 μmol L^{-1} [27–29]. All reagents were dissolved in phosphate buffer saline (PBS) having a pH of 7.2, centrifuged at 2500 g (Centric, Tehtnica D.o.o., Železniki, Slovenia) for 15 minutes, and the supernatants were sterilized by filtration through 0.2 μm syringe filter (Millipore, USA). Sterilized reagents were stored on −20°C until their use in the experiments.

2.2 Cell culture

CaCo-2 cells were cultivated in a complete Eagle's medium supplemented with 10% of fetal calf serum (FCS) (Sigma-Aldrich, Missouri, USA). The cells were multiplied before the experiments, so that their viability and capability for proliferation can be assayed with MTT Cell Proliferation Assay Kit (K299-100) (BioVision, Milpitas, California, USA). In performed tests, positive cells were further cultivated in a 96-well flat microtiter plates (Sterilin, Sigma-Aldrich, Missouri, USA) or in 25 cm^2 flasks (Sterilin, Sigma-Aldrich, Missouri, USA) in 5% CO_2 at 37°C until the monolayer appeared before they can be used for the experiments.

2.3 The studies' design

The experiments were designed as follows: (1) RJ-F (M), HuIFN-αN3, and 10-HDA in the previously stated concentrations were added alone in a volume of 100 μL/well. (2) In various combinations between them (RJ-F (M), HuIFN-αN3, and 10-HDA) for ratio 1:1, 100 + 100 μL/well were added. For ratio 1:2, 66.8 + 133.2 μL/well were added, and for ratio 2:1, 133.2 + 66.8 μL/well were added.

2.4 Antiproliferative (AP) activity

In the AP assay that was performed in a way according to Sugarman et al. [30], single substances RJ-F (M), HuIFN-αN3, and 10-HDA and their combinations in ratios 1:1, 1:2, and 2:1 were added. In the first well in a row of a 96-well microtiter plate, 200 μL of sample volume was added. This was serially transferred per 100 μL from 1:2 to 1:4096. To these, CaCo-2 cells in a concentration of 10^4 cells/well/100 μL in a complete Eagle's medium with 10% FCS were added. Separately, with CaCo-2 cells without being added alone or in their combinations of 1:1, 1:2, and 2:1, substances served as the negative control. The single tested substances RJ-F (M), HuIFN-αN3, and 10-HDA were used as a positive control. Microtiter plates with substances, as single or in combinations 1:1, 1:2, and 2:1, and CaCo-2 cells were incubated for 72 hours at 37°C in a 5% CO_2 atmosphere. Afterward, the supernatants were discarded, and cells were fixed with 100 μL/well of 10% formalin (Sigma-Aldrich, Missouri, USA) in PBS. After 2 hours, the fixative was removed and cells washed twice with the PBS. Then, 2% Rhodamine (Sigma-Aldrich, Missouri, USA) in a volume of 100 μL/well was added for 15 minutes. This was then removed, and the cells were washed twice with the PBS and air-dried. Optical density (OD) at 550 nm was measured on dried microtiter plates on the Synergy HTX Multi-Mode Reader with Gen 5 software (Biotek, Winooski, USA). The AP activities was determined with the well in rows where 50% cell growth inhibition were found. The AP_{50} was calculated for each separate substance (RJ-F (M), HuIFNαN3, and 10-HDA) or their combinations (1:1, 1:2, and 2:1).

2.5 Glutathione (GSH) determination

According to Watson et al. [24], the GSH assay was performed. The CaCo-2 cells were cultivated in flasks having 25 cm^2 (Sterilin, Sigma-Aldrich, Missouri, USA) in a complete Eagles' medium with 10% FCS. To the monolayer of CaCo-2 cells, the substances of RJ-F (M), HuIFN-αN3, and 10-HAD alone in volumes of 1.0 mL, or their combinations of 1:1, 1:2, and 2:1 in volumes of 2.0 mL, were added. The treatment took place for 24 hours at 37°C and 5% CO_2. Then, the cells were removed by Trypsin (Sigma-Aldrich, Missouri, USA) and treated with 1 mL of mmol mL^{-1} Tris–HCl solution (pH = 6.0) containing 0.5 mol mL^{-1} diethylene triamine pentacetic acid (DTPA) (Sigma-Aldrich, Missouri, USA). After such treatment, the cells were lysed by the syringing through the insulin syringe. The quantities of total cellular proteins were determined by the Bio-Rad protein assay (Bio-Rad Laboratories, Hercules, California, USA) having a standard bovine serum albumin (BSA) (Sigma-Aldrich, Missouri, USA).

To determine the amount of GSH, 50 μL of the lysed CaCo-2 cells, 100 μL of 25 μmol L^{-1} DL-Dithiothreitol (DTT), and 150 μL of 0.1 mol mL^{-1} Tris-HCl (pH 8.5) were added. This mixture was incubated for 30 minutes on ice. Then, 750 μL of 2.5% (wt. /vol) 5-sulfosalicylic acid was added to precipitate proteins. The precipitate was centrifuged at 13000 g (Centric, Tehtnica D.o.o., Železniki, Slovenia) for 5 minutes at 4°C, and the supernatants were used in GSH Assay Kit (Sigma-Aldrich, Missouri, USA) to measure the GSH level at OD at 412 nm on Synergy HTX Multi-Mode Reader with Gen 5 software (Biotek, Winooski, USA) and express it as nmol of GSH g^{-3}/proteins.

2.6 Lipid peroxidation assay

CaCo-2 cells were cultivated in 25 cm^2 flasks (Sterilin, Sigma-Aldrich, Missouri, USA) in a complete Eagles' medium with 10% FCS in 5% CO_2 and 37°C. When monolayer was formed, the cells were treated with substances either alone or their combinations as described previously, in a quantity of 1.0 mL when single substances were added, and a total of 2.0 mL when their combinations (1:1, 1:2, and 2:1) were added. The cells were incubated for 24 hours at 37°C and 5% CO_2. The medium was removed, and cells were detached with trypsin, washed, and resuspended in 5 mL of PBS. The total cell protein concentration was determined by the Bio-Rad protein assay (Bio-Rad Laboratories, Hercules, California, USA) with a BSA as a standard. A measure of 1 mL of thiobarbituric acid (TBA) reagent (0.38% 2-TBA, 15.0% TBA, 0.3 mol mL^{-1}HCl) was added to the cell suspension, and these were heated at 95°C for 20 minutes, chilled to room temperature, and centrifuged at 1500 g for 10 minutes (Centric, Tehtnica d.o.o., Železniki, Slovenia). The quantity of TBA reactive substances (RS) developed by lipid peroxidation was measured at OD at 535 nm in Synergy HTX Multi-Mode Reader with Gen 5 software, (Biotek, Winooski, USA). The obtained results were expressed as MDA nmol g^{-3} of protein.

2.7 Statistical evaluation

For statistical significance determination ($^*p < 0.1$, $^{**}p < 0.05$), the t-test was used. All the data are shown as a mean value ± standard deviation. The tests were performed in triplicate, and each was repeated three to four times.

3. Results and discussion

3.1 Antiproliferative (AP) activity

The results of AP activity and their concentrations at AP_{50} in case of single substances were found as follows: RJ-F (M), 0.157 (0.5 mg mL^{-1}); HuIFN-αN3, 0.741 (208.33 IU mL^{-1}; and 10-HDA, 0.487 (37.5 μmol L^{-1}). The AP activities of used single substances (RJ, HuIFN-αN3, and 10-HDA) were relatively low. Such AP activity expressed as growth rate index (GRI) of their combinations (1:1, 1:2, 2:1) is shown in **Figure 1**.

When a combination between RJ-F (M) and HuIFN-αN3 in ratio 2:1 was used, the highest AP activity was found. Their AP activity expressed as GRI was 1.006. The combinations of 10-HDA and HuIFN-αN3 2:1 show the AP activity expressed as GRI 0.952, which was lower than this in combination of RJ-F (M) and HuIFN-αN3 (2:1). These results suggest that RJ-F (M) should contain some other components responsible for the relatively strong AP activity of the combination of the RJ-F (M) and HuIFN-αN3 (2:1). The RJ $protein_{30}$, a water soluble fraction showing the cytotoxic effect on the HeLa cells by decreasing the cell population by 50%, should be taken into account.

The AP activity of RJ-F (M), HuIFN-αN3, and 10-HDA on the CaCo-2 cells is connected to the induction of apoptosis and cytotoxicity. They also influence the glutathione level and lipid peroxidation in CaCo-2 cells [31].

3.2 GSH determination and measurement of lipid peroxidation via MDA

The results obtained in our study show that RJ-F (M), HuIFN-αN3, 10-HDA, and their combinations (1:1, 1:2, and 2:1) decreased the intracellular level of the GSH and increased the lipid peroxidation via the MDA. The data in extent are shown in **Figure 2** and **Table 1.**

The GHS has an important role in various intracellular processes like cellular differentiation, proliferation, apoptosis, and cancer. The decrease of the GSH/glutathione disulfide (GSSG) ratios consequently lead to the enhanced susceptibility to oxidative stress and the progression of cancer. In addition, the increased GSH levels enhance the antioxidant capacity and resistance to oxidative stress that can be found in transformed cells. Some of the possible antitumor mechanisms of RJ-F (M) are linked to the modulation of the oxidative stress and induction of apoptosis [32]. Similar effects were found during the treatment of the PaCa-44 cells with 10-HDA, where induction of the apoptosis was also confirmed [33]. Furthermore, the antitumor activity of the HuIFN-αN3 is connected with the induction of apoptosis and the modulation of the oxidative stress in rats having breast cancer [34].

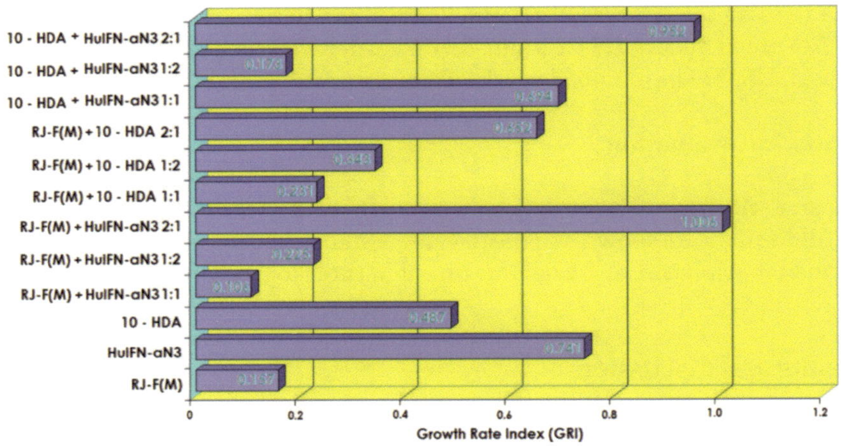

Figure 1.
The antiproliferative (AP) effect expressed as growth rate index (GRI = GR_{in}.−GR_{8d}/GR_{8d}) of RJ-F (M), HuIFN-αN3, 10-HDA, and their combinations: 1:1, 1:2, and 2:1 on the CaCo-2 cells. The inhibition of proliferation of CaCo-2 cells treated with RJ-F (M) (0.1 g/10 mL PBS), HuIFN-αN3 (1000 IU mL^{-1}), 10-HDA (100 μmol mL^{-1}), and different combinations between them (1:1, 1:2, and 2:1) were analyzed. The AP activity was determined by Rhodamine B staining method after 72 hours of incubation at 37°C and 5% CO_2.

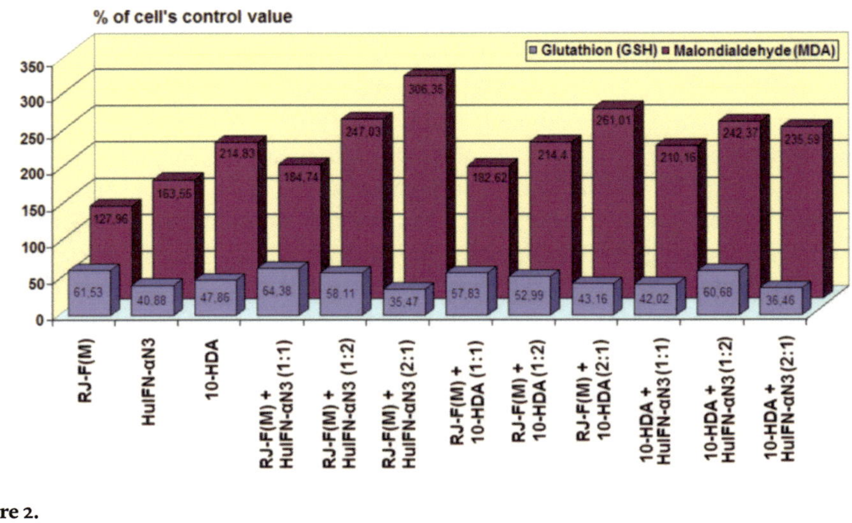

Figure 2.
The percent of CaCo-2 cell control value of glutathione (GSH) or malondialdehyde (MDA) after treatment with RJ-F (M), HuIFN-αN3, 10-HDA, and different combinations between them (1:1, 1:2, and 2:1).

Values for	Cell Control	RJ-F (M) b)	Hu IFN-αN3 c)	10-HDA d)	RJ-F (M) + HuIFN-αN3 (1:1)	RJ-F (M) + HuIFN-αN3 (1:2)	RJ-F (M) + HuIFN-αN3 (2:1)	RJ-F (M) + 10-HDA (1:1)	RJ-F (M) + 10-HDA (1:2)	RJ-F (M) + 10-HDA (2:1)	10-HDA + HuIFN-αN3 (1:1)	10-HDA + HuIFN-αN3 (1:2)	10-HDA + HuIFN-αN3 (2:1)
Glutathion (GSH) (Mean±SE) a)	70.2 ± 3.2	43.8 ± 2.8	28.7 ± 6.4	33.6 ± 5.8	45.2 ± 4.7	40.8 ± 3.1	24.9 ± 2.4	40.6 ± 4.5	37.2 ± 2.1	30.3 ± 3.7	29.5 ± 1.7	42.6 ± 5.3	25.6 ± 3.1
T-test: T		37.33	58.68	51.76	35.35	41.57	64.06	41.86	46.66	56.42	56.14	37.19	64.88
T-test: p		0.00 035	0.00 014	0.00 018	0.00 04	0.00 028	0.00 012	0.00 028	0.00 022	0.00 031	0.00 031	0.00 072	0.00 023
T-test: r		p<.05	p<.05	p<.05	p<.05	p<.05	p<.05	p<.05	p<.05	p<.05	p<.05	p<.05	p<.05
Malondi-aldehyde (MDA) (Mean±SE) a)	23.6 ± 9.1	30.2 ± 4.3	38.6 ± 4.2	50.7 ± 4.6	43.6 ± 4.1	58.3 ± 5.2	72.3 ± 3.1	43.1 ± 2.6	50.6 ± 4.5	61.6 ± 5.2	49.6 ± 4.2	57.2 ± 2.6	55.6 ± 6.2
T-test: T		-7.88	-19.70	-36.72	-31.42	-54.52	-76.52	-26.03	-42.42	-52.06	-50.11	-45.87	-50.28
T-test: p		0.01 572	0.00 256	0.0007 4	0.00 1011	0.00 0336	0.00 0171	0.00 1472	0.00 0555	0.00 0369	0.00 0398	0.00 0475	0.00 0395
T-test: r		p<.05	p<.05	p<.05	p<.05	p<.05	p<.05	p<.05	p<.05	p<.05	p<.05	p<.05	p<.05

Table 1.
Glutathione (GSH) determination and the measurement of lipid peroxidation (MDA) after the CaCo-2 cell treatment with RJ-F(M), HuIFN-αN3, 10-HDA, and their combinations of 1:1, 1:2, and 2:1.

The 10-HDA, the most important constituent of the RJ-F (M) and HuIFN-αN3, had a complementary active point due to its antitumor activity, which can be enhanced when they were combined in a proper ratio. The most effective was the ratio of 2:1 of the RJ-F (M) and HuIFN-αN3. Using this, the level of GSH was 24.9 ± 2.4 nmol g^{-3}, and the level of MDA was 72.3 ± 3.1 nmol g^{-3}. The following were the control values: for GSH it was 70.2 ± 3.2 nmol g^{-3} and for MDA it was 23.6 ± 9.1 nmol g^{-3}. The RJ-F (M) and HuIFN-αN3 applied at 2:1 ratio and HuIFN-αN3 applied in combination with 10-HDA in 2:1 ratio caused a decrease in the level of GSH and increase in lipid peroxidation indicator level (MDA) in CaCo-2 cells in comparison with the control group. On the other hand, it was observed that these combinations had the highest antiproliferative effect. In addition, it is suggested that antiproliferative effects of RJ-F (M), HuIFN-αN3, and 10-HDA on the CaCo-2 cells can be connected not only with the induction of apoptosis and cytotoxicity but also with their influence on the pro- and anti-oxidative balance [35].

4. Conclusion

Future experiments will show whether these GSH- and MDA-related activities of the RJ-F (M), HuIFN-αN3, 10-HDA, and their combinations may cause the decrease of the tumorigenicity index of different tumor cells in vitro as a result of their tumorigenic potential, as it has already been reported in the literature [36–39]. This is important for practical use of the RJ-F (M), 10-HDA, and HuIFN-αN3 in a proper combination that could be of value for future development of possible tumor therapy based on the use of these bioactive compounds.

Acknowledgements

This research was performed via the Project "Oncolytic Newcastle disease virus in the veterinary medicine" of the Croatian Institute for Experimental and Translational Medicine, Koledinečka 03, 10040 Zagreb, Croatia. Authors are indebted to Tomaž Velnar for English suggestions.

Conflict of interests

No conflicts of interest were declared by all authors.

Author details

Bratko Filipič[1]*, Lidija Gradišnik[2], Klemen Rihar[3†], Adriana Pereyra[4], Damir Đermić[5] and Hrvoje Mazija[1]

1 Croatian Institute for Experimental and Translational Oncology (CIETO), Zagreb, Croatia

2 Medical Faculty of Maribor, Institute of Biomedical Sciences, Maribor, Slovenia

3 Chengdujska, Ljubljana, Slovenia

4 Medex D.o.o., Ljubljana, Slovenia

5 Division of Molecular Biology, Ruđer Bošković Institute, Zagreb, Croatia

*Address all correspondence to: bratko.filipic@gmail.com

†Paper is devoted to the remembrance of Dr. Klemen Rihar, who passed away in December 2018.

References

[1] Oršolić N. Royal jelly: Components, efficiency, analysis and standardization. Arhiv za Higijenu Rada i Toksikologiju. 2013;**64**:445-461. DOI: 10.2478/10004-1254-64-2013-2332

[2] Buttstedt A, Moritz RFA, Bobiş O, Vezeteu TV. Food to some, poison to others honeybee royal jelly and its growth inhibiting effect on European foulbrood bacteria. MicrobiologyOpen. 2017;**6**:e00397. DOI: 10.1002/mbo3.397

[3] Sugiyama T, Takahashi K, Mori H. Royal jelly acid, 10-hydroxy-trans-2-decenoic acid as a modulator of the innate immune responses. Endocrine, Metabolic & Immune Disorders-Drug Targets. 2012;**12**:368-376. DOI: 10.2174/187153012803832530

[4] Honda Y, Araki Y, Hata T, Ichihara K, Ito M, Tanaka M, et al. 10-Hydroxy-2-decenoic acid, the major lipid component of royal jelly, extends the lifespan of *Caenorhabditis elegans* through dietary restriction and target of rapamycin signaling. Journal of Aging Research. 2015;**2015**:425261. DOI: 10.1155/2015/425261

[5] Miyata Y, Sakai H. Anti-cancer and protective effects of royal jelly for therapy-induced toxicities in malignancies. International Journal of Molecular Sciences. 2018;**19**:3270-3273. DOI: 10.3390/ijms19103270

[6] Šver L, Oršolić N, Tadić Z, Njari B, Valpotić I, Bašić I. A Royal Jelly as a new potential Immunomodulator in rats and mice. Comparative Immunology, Microbiology & Infectious Diseases. 1996;**19**:31-38. DOI: 10.1016/0147-9571(95)00020-8

[7] Majtan J, Kumar P, Majtan T, Walls AF, Klaudiny JW. Effect of honey and its major royal jelly protein 1 on cytokine and MMP-9 mRNA transcripts in human keratinocytes. Experimental Dermatology. 2010;**19**:e73-e79. DOI: 10.1111/j.1600-0625.2009.00994.x

[8] Ramadan FM, Al-Ghamdi A. Bioactive compounds and health-promoting properties of royal jelly: A review. Journal of Functional Foods. 2012;**4**:39-52. DOI: 10.1016/j.jff.2011.12.007

[9] Buttstedt A, Moritz RFA, Erler SM. Origin and function of the major royal jelly proteins of the honeybee (*Apis mellifera*) as members of the yellow gene family. Biological Reviews. 2013;**89**:255-269. DOI: 10.1111/brv.12052

[10] Koomankode NA, Ramanathana G, Naira JA, Sugunanb VS. A review on Royal Jelly proteins and peptides. Journal of Functional Foods. 2018;**44**:255-264. DOI: 10.1016/j.jff.2018.03.008

[11] Corona M, Velarde RA, Remolina S, Moran-Lauter A, Wang Y, Hughes KA, et al. Vitellogenin, juvenile hormone, insulin signaling, and queen honey bee longevity. PNAS. 2007;**104**:7128-7133. DOI: 10.1073/pnas.07019099104

[12] Khazaei M, Ansarian A, Ghanbari E. New findings on biological actions and clinical applications of Royal Jelly: A review. Journal of Dietary Supplements. 2017;**15**:757-775. DOI: 10.1080/19390211.2017.1363843

[13] Tarhini AA, Gogas H, Kirkwood JM. IFN-alpha in the treatment of melanoma. Journal of Immunology. 2012;**189**:3789-3793. DOI: 10.4049/jimmunol.1290060

[14] Crow MK. Type I interferon in organ-targeted autoimmune and inflammatory diseases. Arthritis Research & Therapy. 2010;**12**(Suppl 1): S1-S10. DOI: 10.1186/ar2886

[15] Maedaa S, Wadab H, Naitoa Y, Naganob H, Simmonsa S, Kagawaa Y, et al. Interferon-α acts on the

S/G2/M phases to induce apoptosis in the G1 phase of an IFNAR2- expressing hepatocellular carcinoma cell line. The Journal of Biological Chemistry. 2014;**289**:23786-23795. DOI: 10.1074/jbc.M114.551879

[16] Zitvogel L, Galluzzi L, Kepp O, Smyth MJ, Kroemer G. Type I interferons in anticancer immunity. Nature Reviews Immunology. 2015;**15**:405-414. DOI: 10.1038/nri3845

[17] Diamond MS, Kinder M, Matsushita H, Mashayekhi M, Dunn GP, Archambault JM, et al. Type I interferon is selectively required by dendritic cells for immune rejection of tumours. Journal of Experimental Medicine. 2011;**208**: 1998-2003. DOI: 10.1084/jem.20101158

[18] Qu J, Hou Z, Han Q, Zhang C, Tian Z, Zhang J. Poly(I:C) exhibits an anti-cancer effect in human gastric adenocarcinoma cells which is dependent on RLRs. International Immunopharmacology. 2013;**17**: 814-820. DOI: 10.1016/j.intimp.2013.08.013

[19] Smakman N, van den Wollenberg DJM, Borel Rinkes IHM, Hoeben RC, Kranenburg O. Sensitization to apoptosis underlies KrasD12-dependent oncolysis of murine C26 colorectal carcinoma cells by reovirus T3D. Journal of Virology. 2005;**79**:14981-14985. DOI: 10.1128/JVI.79.23.14981-14985.2005

[20] Anguille S, Lion E, Willemen Y, Van Tendeloo VFI, Berneman ZN, Smits ELJM. Interferon-α in acute myeloid leukemia: An old drug revisited. Leukemia. 2011;**25**:739-748. DOI: 10.1038/leu.2010.324

[21] Junko K, Yumiko N, Satoru K, Motoshi T, Misao O. Effects of interferon-τ on cattle persistently infected with bovine viral diarrhea virus. Japanese Journal of Veterinary Research. 2012;**60**:63-70. DOI: 10.14943/jjvr.60.2-3.63

[22] Kohara J, Nishikura Y, Tajima M, Onuma M, Yokomizo Y. Antiviral effects of bovine IFN- τ and human IFN- α on bovine viral diarrhea virus. Veterinary Immunology and Immunopathology. 2009;**128**:211-347. DOI: 10.1016/j.vetimm.2008.10.258

[23] Jonas CR, Ziegler TR, Gu LH, Jones DP. Extracellular thiol/disulfide redox state affects proliferation rate in a human colon carcinoma (Caco2) cell line. Free Radical Biology and Medicine. 2002;**33**:1499-1506. DOI: 10.1016/s0891-5849(02)01081-x

[24] Watson WH, Yang X, Choi EY, Jones PD, Kehrer PJ. Thioredoxin and its role in toxicology. Toxicological Sciences. 2004;**78**:3-14. DOI: 10.1093/toxsci/kfh050

[25] Niki E. Lipid peroxidation: Physiological levels and dual biological effects. Free Radical Biology and Medicine. 2009;**47**:469-484. DOI: 10.1016/j.freeradbiomed.2009.05.032

[26] Sica G, Fabbroni L, Castagnetta L, Cacciatore M, Pavone-Macaluso M. Antiproliferative effect of interferons on human prostate carcinoma cell lines. Urological Research. 1989;**17**:111-115. DOI: 10.1007/bf00262031

[27] Hidaka S, Okamoto Y, Uchiyama S, Nakatsuma A, Hashimoto K, Onishi S, et al. Royal Jelly prevents osteoporosis in rats: Beneficial effects in ovariectomy model and in bone tissue culture model. Evidence-Based Complementary and Alternative Medicine. 2006;**3**:339-348. DOI: 10.1093/ecam/nel019

[28] Sugiyama T, Takahashi K, Kuzumaki A, Tokoro S, Neri P, Mori H. Inhibitory mechanism of 10-hydroxy-trans-2-decenoic acid (Royal Jelly Acid) against lipopolysaccharide and interferon-β-induced nitric oxide production. Inflammation. 2012;**36**:372-378. DOI: 10.1007/s10753-012-9556-0

[29] Ohkawa H, Ohishi N, Yagi K. Assay for lipid peroxides in animal tissues by thiobarbituric acid reaction. Analytical Biochemistry. 1979;**95**:351-358. DOI: 10.1016/0003-2697(79)90738-3

[30] Sugarman BJ, Aggarwal BB, Hass PE, Figari IS, Palladino MA Jr, Shepard HM. Recombinant human tumor necrosis factor-alpha: Effects on proliferation of normal and transformed cells in vitro. Science. 1985;**230**:943-945. DOI: 10.1126/science.3933111

[31] Estrela JM, Ortega A, Obrador E. Glutathione in cancer biology and therapy. Critical Reviews in Clinical Laboratory Sciences. 2006;**43**:143-181. DOI: 10.1080/10408360500523878

[32] Ibrahim A, Eldaim MAA, Abdel-Daim MM. Nephroprotective effect of bee honey and royal jelly against subchronic cisplatin toxicity in rats. Cytotechnology. 2015;**68**:1039-1048. DOI: 10.1007/s10616-015-9860-2

[33] Kocot J, Kiełczykowska M, Luchowska-Kocot D, Kurzepa J, Musik I. Antioxidant potential of propolis, bee pollen, and royal jelly: Possible medical application. Oxidative Medicine and Cellular Longevity. 2018;**2018**:7074209. DOI: 10.1155/2018/7074209

[34] Park M, Kim H. Anti-cancer mechanism of docosahexaenoic acid in pancreatic carcinogenesis: A mini-review. Journal of Cancer Prevention. 2017;**22**:1-5. DOI: 10.15430/JCP.2017.22.1

[35] Gasic S, Vucevic D, Vasilijic S, Antunovic M, Chinou I, Colic M. Evaluation of the immunomodulatory activities of royal jelly components in vitro. Immunopharmacology and Immunotoxicology. 2007;**29**:521-536. DOI: 10.1080/08923970701690977

[36] Boynton AL, Whitfield JF. Cell biology different calcium requirements for proliferation of conditionally and unconditionally tumorigenic mouse cells. Proceedings of the National Academy of Sciences of the United States of America. 1976;**73**:1651-1654. DOI: 10.1073/pnas.73.5.1651

[37] Parsons PG. Selective proliferation of human tumour cells in calcium—Depleted medium. Immunology and Cell Biology. 1978;**56**:297-300. DOI: 10.1038/icb.1978.31

[38] Whitfield JF, MacManus JP, Rixon RH, Boynton AL, Youdale T, Swierenga S. The positive control of cell proliferation by the interplay of calcium ions and cyclic nucleotides. A review. In Vitro. 1976;**12**:1-18. DOI: 10.1007/bf02832787

[39] Boynton AL, Kleine LP, Whitfield FJ. Relation between colony formation in Calciumdash deficient medium, colony formation in soft agar, and tumor formation by T51B rat liver cells. Cancer Letters. 1984;**21**:293-302. DOI: 10.1016/0304-3835(84)90008-9

Chapter 6

Statins Alone or in Combination with Ezetimibe or PCSK9 Inhibitors in Atherosclerotic Cardiovascular Disease Protection

Marija Vavlukis and Ana Vavlukis

Abstract

Statins have proved to be very effective in reducing atherosclerotic cardiovascular disease (ASCVD) risk, with no apparent threshold at which low-density lipoprotein cholesterol (LDL-C) lowering is not associated with a reduced risk. Yet, several meta-analyses of statin trials show significant on-treatment residual risk of major cardiovascular (CV) events. This finding points to the unmet needs, in terms of LDL-C targets and ASCVD protection, of statin-treated patients, raising the question of statin combination therapy. Ezetimibe is a cholesterol absorption inhibitor, with the potency to decrease LDL-C for about 10–18%, apolipoprotein B (apoB) for 11–16%, while, in combination therapy with statins, leads to an additional LDL-C lowering of 25%, with a total LDL-C lowering of 34–61%. It is also estimated that 10–20% of patients on statin treatment cannot tolerate them. As a result, adequate doses to achieve treatment target, or as recommended for the patient-specific risk profile, cannot be prescribed. Proprotein convertase subtilisin/kexin type 9 (PCSK9) Inhibitors are monoclonal antibodies that inhibit the binding of PCSK9 to LDL-C receptors. Besides a very potent lipid-lowering effect, PCSK9 inhibitors have added ASCVD risk reduction benefit due to a very aggressive LDL-C lowering action, especially beneficial in patients who are intolerant to statins.

Keywords: hyperlipidemia, statins, ezetimibe, PCSK9 inhibitors, cardiovascular diseases, cerebrovascular disease, prevention, adverse effects, diabetes, residual risk

1. Introduction

Atherosclerotic cardiovascular disease (ASCVD), including its clinical manifestations, such as myocardial infarction (MI) and ischemic stroke (IS), is the leading morbidity and/or mortality cause worldwide. One of the most highly studied factors associated with ASCVD is low-density lipoprotein (LDL). Vast evidence has postulated that cholesterol-rich LDL and other apolipoprotein B (apoB)-containing lipoproteins (very low-density lipoproteins (VLDL), intermediate density lipoproteins (IDL), and lipoprotein(a) [Lp(a)]), are directly involved in the development of ASCVD [1].

Statins are the first-line anti-lipemic pharmacotherapy, having been shown to reduce both LDL-C levels and cardiovascular (CV) events. However, a considerable number of statin-treated patients do not achieve target LDL-C levels, even after maximal statin dose-treatment, or are intolerant to intensive statin therapy [2].

In the aforementioned situations patients can largely benefit from an additional LDL-C lowering agent. Ezetimibe is a non-statin drug that can additionally reduce ASCVD risk, when added to a statin, leading to a total of 34–61% LDL-C reduction [3]. Proprotein Convertase Subtilisin-Like/Kexin Type 9 (PCSK9) inhibitors, one of the newest anti-lipemic agents, can lower LDL-C by 45–65%, and are also proven to have ASCVD risk reduction properties [4].

Therefore, the aim of this chapter is to address the question of therapeutic efficacy, as expressed through the lipid-lowering and anti-inflammatory effects, and atherosclerotic cardiovascular disease risk reduction when adding ezetimibe or PCSK9 inhibitors to statin therapy.

1.1 HMG-CoA reductase inhibitors—statins

3-Hydroxy-3-methylglutaryl coenzyme A (HMG-CoA) reductase inhibitors, commonly known as statins, have been one of the most frequently prescribed medications worldwide, since their introduction 30 years ago. Currently, there are six statin drugs available on the market—pitavastatin, atorvastatin, rosuvastatin, pravastatin, simvastatin and fluvastatin [5].

Statins are competitive, reversible inhibitors of HMG-CoA reductase, a rate-limiting step in the process of cholesterol biosynthesis. HMG-CoA is a microsomal enzyme—reductase which catalyzes the conversion of HMG-CoA to L-mevalonate and coenzyme A. Inhibiting the HMG-CoA reductase, statins ultimately prevent the endogenous cholesterol production. Cholesterol concentration reduction triggers an up-regulation of the expression of low-density lipoprotein (LDL)-receptors in the hepatocytes, promoting uptake of LDL and LDL-precursors from the systemic circulation. Therefore, a significant part of the statins' cholesterol-lowering action is a result of an indirect increase in LDL clearance from plasma. Additional mechanisms of action include inhibition of the hepatic apolipoprotein B100 synthesis, and a reduction of the synthesis and secretion of triglyceride-rich lipoproteins [6].

Statins are composed of two parts, the pharmacophore, a dihydroxyheptanoic acid segment, and a moiety composed of a ring system with various substituents. According to the chemical modification of the ring system and the nature of its substituents, different statin structures are generated. Ring substituents define the solubility of the statins, along with many of their pharmacological properties. Among the statins, lovastatin, simvastatin, atorvastatin, and fluvastatin are lipophilic, whereas pravastatin and rosuvastatin are more hydrophilic. Different chemical structures lead to different pharmacokinetic properties, pharmacological effects and pleiotropic actions [7].

Statins can enter the systemic circulation passively, through the intestinal cells, and actively via the ATP-binding cassette (ABC) and solute carrier (SLC) transporters. Two enzyme groups are involved in statin metabolism, the cytochrome P450 (CYP450), and UDP-glucuronosyltransferase (UGT), mainly acting in the liver, and to a lesser extent, in the kidneys. Lipophilic statins are transported via passive diffusion, metabolized by the CYP450 enzymes, and mainly excreted through the biliary system. Hydrophilic statins enter the liver via active transport, and are actively excreted through the kidneys, mostly as unchanged drugs. Lipophilic statins have generally low bioavailability due to first pass metabolism. Absorption varies between 30 and 98%, and time to reach peak plasma concentration (Tmax) is within 4 h of administration. Statins are administrated orally as active hydroxy acids, except for lovastatin and simvastatin, which are administrated as lactone prodrugs, and then hydrolyzed to the hydroxy acid form. Their bioavailability varies;

pitavastatin has a bioavailability of 80%, whereas fluvastatin between 19 and 29%. The CYP3A4 isoenzyme is responsible for the metabolism of lovastatin, simvastatin and atorvastatin. Their active metabolites—2-hydroxy- and 4-hydroxy-atorvastatin acid from atorvastatin, and β-hydroxy simvastatin acid from simvastatin, carry a part of their inhibitory activity. Fluvastatin is mainly metabolized by the CYP2C9 isoenzyme. Pravastatin is eliminated by both the kidney and liver, mostly as an unchanged drug [6–8].

1.2 Cholesterol absorption inhibitors—ezetimibe

Ezetimibe, a cholesterol absorption blocker, has been the focus of many trials supporting its use in ASCVD risk reduction. For patients that cannot achieve target treatment goals with statin therapy alone, ezetimibe has proven to be a safe, well-tolerated medication which, combined with statins, leads to additional LDL-C reduction, thus resulting in a significant morbidity and/or mortality benefit [9].

Serum cholesterol is derived from two major sources: cholesterol synthesized *de novo* in the liver and cholesterol that has been absorbed from the gastrointestinal tract. Statins reduce serum cholesterol by reducing its biosynthesis in the liver. Ezetimibe, on the other hand, targets gastrointestinal cholesterol absorption. Ezetimibe acts at the brush boarder of the small intestine, by selectively inhibiting the cholesterol transport protein Niemann Pick C1 like 1 protein (NPC1L1), thus preventing uptake of intestinal luminal cholesterol micelles into the enterocytes. The reduced cholesterol uptake leads to hepatic LDL-C stores depletion, resulting in upregulation of hepatic LDL receptors, causing LDL-C clearance from the blood. It is also suggested that ezetimibe inhibits the hepatic NPC1L1 as well, thus leading to reduced hepatic cholesterol absorption [3, 10, 11].

Following ingestion, the drug is extensively (>80%) metabolized to its active form—ezetimibe-glucuronide. Glucuronidation of the 4-hydroxyphenyl group, by uridine 5′-diphosphate-glucuronosyltransferase isoenzymes, forms the major ezetimibe metabolite in the intestine and liver. Total ezetimibe (sum of 'parent' ezetimibe plus ezetimibe-glucuronide) concentrations reach a maximum 1–2 h after administration. Both the parent compound and the glucuronidated compound are absorbed, and recirculated via the hepatobiliary excretion, thus providing long-term cholesterol absorption inhibition. This cycle accounts for the long half-life of ezetimibe—about 22 h, allowing for once-a-day dosing. About 10–15% of the drug is excreted in the urine, and the rest in the feces, mainly as the parent drug. Ezetimibe does not appear to be metabolized or interact with the cytochrome P450 pathway, thus it does not affect bioavailability and kinetics of commonly used drugs that are affected by the CYP450 family [10, 12].

1.3 Proprotein convertase subtilisin/kexin type 9 (PCSK9) inhibitors

The discovery of PCSK9 in 2003 opened many new research directions in the cardiovascular field. Liver PCSK9 binds to the LDL receptor (LDL-R) and promotes its degradation in the endosomal/lysosomal pathway. Higher PCSK9 activity leads to lower liver LDL-R levels, resulting in reduced LDL-uptake from circulation, and thus in hypercholesterolemia [13].

This led to a conclusion that the inhibition of PCSK9 would mean that more LDL receptors would be recycled to the surface of the cell, thus increasing the clearance of LDL cholesterol from the circulation. Since then various approaches to the pharmacological inhibition of PCSK9 have been investigated, and parenteral anti-PCSK9 monoclonal antibodies (MoAbs) have been the most successful strategy to date. MoAbs are now in late-stage (phase 3 clinical trials) testing [14].

Anti-PCSK9 MoAbs are known to bind at or near PCSK9's binding site for the LDL-R. This sterically inhibits the interaction of PCSK9 with the LDL-R, thus reducing the degradation of the receptor. This markedly increases the clearance of LDL and substantially lowers plasma LDL cholesterol, as well as apolipoprotein-B100 [15]. In 2015, the Food and Drug Administration (FDA) approved the PCSK9 inhibitors alirocumab and evolocumab for patients with clinical atherosclerotic cardiovascular disease on maximally tolerated statin therapy who "require additional lowering of LDL-C" [16].

Evolocumab is a human monoclonal immunoglobulin G2 antibody directed against the circulating PCSK9 protein. Evolocumab is administered by subcutaneous injection to the abdomen, thigh, or upper arm. For patients with primary hyperlipidemia, who have clinical ASCVD, or heterozygous familial hypercholesterolemia, the recommended subcutaneous dose is 140 mg every 2 weeks or 420 mg once monthly. Maximum suppression of circulating unbound PCSK9 is seen after 4 h. Peak serum concentrations are obtained in 3–4 days, with an estimated bioavailability of 72%. The drug is estimated to have an effective half-life of 11–17 days [17].

Alirocumab is a human monoclonal immunoglobulin G1 [IgG1] isotype antibody that binds to circulating PCSK9, thus inhibiting its action on LDL-R. Alirocumab reduces free PCSK9 in a concentration-dependent manner. Following a single subcutaneous administration of alirocumab 75 or 150 mg, maximal suppression of free PCSK9 occurs within 4–8 h. Within 4–8 weeks after initiating or titrating alirocumab therapy, LDL-C levels should be tested to determine the response and the need for (additional) dose adjustments. The drugs' median apparent half-life at steady state is 17–20 days. Peak serum concentrations are obtained in 3–7 days, with an estimated bioavailability of 85%. At low concentrations, the elimination of alirocumab occurs predominately via saturable binding to PCSK9. At higher concentrations, elimination is through a nonsaturable proteolytic pathway [18].

2. Effects of statins

2.1 Effects of statin therapy on LDL-C concentrations

During the past 20 years, the extensive use of statin therapy among patients known to have an occlusive vascular disease, or are considered to be at increased risk of cardiovascular events, has been associated with descending actions on LDL and total cholesterol concentrations [19].

Different statins have different potencies, with the newer agents (e.g., atorvastatin and rosuvastatin) able to produce larger reductions in LDL cholesterol per mg of drug, compared to the older agents (e.g., simvastatin and pravastatin). Each dose doubling leads to an additional reduction of about 6 percentage points in LDL cholesterol (e.g., 43 vs. 49% reductions with atorvastatin 20 vs. 40 mg daily). The American College of Cardiology/American Heart Association (ACC/AHA) 2013 Blood Cholesterol Guideline classified statin regimens as being of low intensity (e.g., $<30\%$ LDL-C reduction with simvastatin 10 mg daily), moderate intensity (e.g., 30% to $<50\%$ reduction with simvastatin 20–40 mg, atorvastatin 10–20 mg, or rosuvastatin 5–10 mg daily), or high intensity (e.g., $\geq 50\%$ reduction with atorvastatin 40–80 mg or rosuvastatin 20–40 mg daily) [20].

High-intensity statin therapy would be expected to reduce LDL-C by at least 2 mmol/L in individuals with LDL-C concentrations of 4 mmol/L or more, but by only about 1 mmol/L in those presenting with concentrations of 2 mmol/L. Consequently, since vascular events rates reductions, in patients treated with statins, are related to the absolute reductions in LDL-C, intensive statin treatment should be used in individuals at higher risk of vascular events, rather than just on those with high cholesterol concentrations [21].

The Cholesterol Treatment Trialists' (CTT) Collaboration was settled to conduct meta-analyses of randomized controlled statin-oriented trials involving at least 2 years of treatment in at least 1000 patients. During the study treatment periods (on average 5 years), the average LDL-C reduction was about 1–1.5 mmol/L, comparing routine statin therapy vs. no routine statin therapy, with an additional LDL-C reduction of about 0.5 mmol/L in the trials comparing allocation to more vs. less intensive statin regimens. To summarize, an intensive statin regimen, compared to no statin therapy, reduced LDL-C concentrations by 1.5–2 mmol/L [22, 23].

2.2 Reductions in major vascular event (MVE) rates

Statins have been proven to be very effective in reducing ASCVD risk, with no apparent threshold at which LDL-C lowering is not associated with reduced risk. The Atherosclerosis Risk in Communities (ARIC) study, performed on 13,342 individuals, provided evidence that protection against ASCVD happens in a graded fashion with LDL-C level [24]. The CTT meta-analyses detected about 25,000 major vascular events (MVE) (composite of coronary deaths or non-fatal myocardial infarctions, strokes of any type, and coronary revascularisation procedures). Comparing routine vs. no routine statin treatment, there was a 20% proportional reduction in the MVE rate per mmol/L LDL-C reduction. Regarding the comparison of more vs. less intensive statin regimens, the average 0.5 mmol/L further LDL-C reduction lead to an additional 15% proportional reduction in the MVE rate [22, 23].

By combining the findings from the two previously mentioned sets of trials, it can be concluded that a LDL-C concentration reduction by 2 mmol/L would reduce the MVE risk by about 45%. Given the aforementioned, larger LDL-C reductions should lead to larger risk reductions (e.g., 60–70% with 3–4 mmol/L LDL-C reductions); however, this is likely only to be clinically relevant in limited circumstances (such as for individuals with familial hypercholesterolemia who have very high LDL-C levels) [25].

High-intensity statin treatment (atorvastatin 80 mg) in the Treating to New Targets (TNT), the Incremental Decrease in Endpoints Through Aggressive Lipid Lowering (IDEAL) trial and Pravastatin or Atorvastatin Evaluation and Infection Therapy (PROVE-IT) trial, demonstrated an additional 11–23% relative risk reduction of major CVD events, when compared to moderate-intensity statin therapy (atorvastatin 10 mg, simvastatin 20–40 mg, or pravastatin 40 mg). Nonetheless, the atorvastatin 80 mg treated patients still experienced a major CVD event during the trials (ranging from 4 to 11% per year). Mean LDL-C levels in the atorvastatin 80-mg groups ranged from 1.6 to 2.1 mmol/L [26–29].

The American College of Cardiology/American Heart Association 2013 Blood Cholesterol Guideline gives recommendations regarding statin therapy in terms of ASCVD prevention and risk reduction (**Table 1**) [20].

2.3 Reductions in coronary mortality

The CTT meta-analyses showed a 12% proportional reduction in vascular mortality per mmol/L LDL-C reduction, attributable to an approximately 20% proportional reduction in coronary deaths, 8% reduction in other cardiac deaths, and little effect on death due to all types of stroke combined. No matter the cause of coronary death, the risk reduction per mmol/L LDL-C reduction appear to be similar in patients with and without pre-existing vascular disease, and in those who present at different levels of baseline vascular risk [22, 23].

ASCVD Statin Benefit Groups [20] Heart healthy lifestyle habits are the foundation of ASCVD prevention			
Clinical ASCVD	**LDL-C ≥ 4.9 mmol/L**	**Diabetes; age 40–75 years[1]**	**Estimated 10-yr ASCVD risk ≥ 7.5%[2]; age 40–75 years[1]**
• High-Intensity statin (age ≤ 75 years) • Moderate-intensity statin if >75 years or not a candidate for high-intensity statin	• High-intensity statin • Moderate-intensity statin if not a candidate for high-intensity statin	• Moderate-intensity statin • High-intensity statin if estimated 10 year ASCVD risk ≥ 7.5%	• Moderate- to high-intensity statin

ASCVD, atherosclerotic cardiovascular disease; LDL-C, low-density lipoprotein-cholesterol.
[1]With LDL-C of 1.8–4.9 mmol/L.
[2]Estimated using the Pooled Cohort Risk Assessment Equations.

Table 1.
Statin therapy in ASCVD prevention and risk reduction according to the 2013 ACC/AHA blood cholesterol guideline.

Regarding the effect of different statins, and different statin treatment intensities, on coronary mortality, the TNT trial showed no significant differences in the risk of death from cardiovascular or noncardiovascular causes between the patients treated with 10 mg or 80 mg atorvastatin per day [27]. The IDEAL study compared the effects of high-intensity statin therapy (atorvastatin 80 mg/d) vs. low-intensity statin therapy (usual-dose simvastatin, 20 mg/d), on occurrence rates of a major coronary event, defined as coronary death, confirmed nonfatal acute MI, or cardiac arrest with resuscitation. The results failed to show a statistically significant difference in all-cause or cardiovascular mortality between the two treatment regimens [28]. The PROVE-IT trial aimed to compare the effects of 40 mg of pravastatin daily (standard therapy) vs. 80 mg of atorvastatin daily (intensive therapy) in patients hospitalized for acute coronary syndrome. The risk of death due to coronary heart disease, myocardial infarction, or revascularization was reduced by 14% in the atorvastatin group, as compared with 22.3% in the pravastatin group [29].

2.4 The question of residual risk

Despite what was previously elaborated, a significant on-statin treatment residual risk of major CV events still exists. A meta-analysis of statin trials shows that there is residual CVD event risk even with LDL-C levels <2 mmol/L. The aforementioned TNT trial, conducted on patients with stable coronary artery disease (CAD), described an 8.7% incidence of a major event, over 5 years, in patients receiving 80 mg atorvastatin daily, with on-treatment LDL-C concentrations of 1.8–2.6 mmol/L [24]. Findings like these point to the unmet needs of the patients treated with statins. Several cholesterol treatment guidelines recommend a LDL-C treatment goal of <2.6 mmol/L or < 1.8 mmol/L, depending on the level of risk. However, in the everyday clinical practise many high-risk patients fail to reach the goal [26].

The most resent Guidelines, the 2016 European Society of Cardiology and European Atherosclerosis Society (ESC/EAS) Guidelines for the management of Dyslipidemias and The 2017 Guidelines of the American Association of Clinical Endocrinologists (AACE) and the American College of Endocrinology (ACE) for Management of Dyslipidemia and Prevention of Cardiovascular Disease have recommended similar target LDL-C levels, and have suggested the use of combination therapy (ezetimibe and PCSK9 inhibitors) to achieve these targets in situations in which maximally tolerated statin monotherapy is insufficient (**Table 2**) [3, 26].

The AACE guidelines introduced an additional "extreme high-risk" category, which is not recognized by the ESC/EAS, and an additional treatment LDL-C target

Risk profile of the patient	Treatment target goal of LDL-C [3]
Very high CV risk	<1.8 mmol/L, or at least 50% reduction if the baseline[3] LDL-C is 1.8–3.5 mmol/L (COR[1] I/LOE[2] B)
High CV risk	<2.6 mmol/L, or at least 50% reduction if the baseline LDL-C is 2.6–5.2 mmol/L (COR I/LOE B)
Low/moderate CV risk	LDL-C goal of <3.0 mmol/L (COR IIa/LOE C)

CV, cardiovascular; LDL-C, low-density lipoprotein-cholesterol.
[1]COR, class of recommendation.
[2]LOE, level of evidence.
[3]"Baseline LDL-C" refers to the LDL-C level in a subject not taking any lipid lowering medication.

Table 2.
Treatment target goal of LDL-C according to the 2016 ESC/EAS guidelines for the management of dyslipidemias.

of <1.4 mmol/L. This "extreme high-risk" group represents patients with progressive disease, despite LDL-C levels of <1.8 mmol/L while on-statin therapy. The rationale of the aforementioned approach is in the individualization of the total CV risk reduction, which can be better done if goals are predefined. Treatment goals are defined and tailored to the total CV risk level of each individual patient. The "individualized approach "may possibly result with better patient adherence to the therapy. The growing number of evidence suggests that LDL-C lowering beyond the guidelines-set goals may lead to further reduction of CVD events, which can be especially beneficial in patients at very high CV risk [3, 30].

Given what was previously discussed, in order to achieve this level of LDL-C reduction, combination therapy may be needed. The latest randomized clinical trials (RCTs), such as The LDL-C Assessment With Proprotein Convertase Subtilisin Kexin Type 9 Monoclonal Antibody Inhibition Combined With Statin Therapy 2 (LAPLACE-2) trial, The FOURIER (Further Cardiovascular Outcomes Research With PCSK9 Inhibition in Subjects With Elevated Risk) trial, and The Reductions in Atherogenic Lipids and Major Cardiovascular Events: A Pooled Analysis of 10 ODYSSEY Trials Comparing Alirocumab With Control, demonstrated that extremely low LDL-C levels (<0.5 mmol/L) appear to be safe. Furthermore, The IMPROVE-IT (Examining Outcomes in Subjects With Acute Coronary Syndrome: Vytorin vs. Simvastatin) and FOURIER trials demonstrated that not only such low levels are safe, but are also beneficial, in terms of additional CV risk reduction [30–34].

3. Combination therapy

3.1 Combination therapy: ezetimibe ad-on statin

The unmet needs in terms of LDL-C targets and ASCVD protection raised the question of statin combination therapy. It only needed to be right positioned. Such positioning was done in the 2016 ESC/EAS Guidelines for the management of Dyslipidemias (**Table 3**) [3], and also in The 2016 ACC expert consensus decision pathway on the role of non-statin therapies for LDL-cholesterol lowering in the management of atherosclerotic cardiovascular disease risk [35].

3.1.1 Effects of ezetimibe ad-on statins on LDL-C

The FDA-approved ezetimibe indications are for treatment of primary hyperlipidemia, alone or in combination with a statin; mixed hyperlipidemia in combination

Clinical setting	COR[1] LOE[2]	Treatment target goal of LDL-C	COR[1] LOE[2]
Hypercholesterolemia If the goal is not reached with statins, ad-on ezetimibe	IIa/B	Depends of the risk profile of the individual patient	I/A
FH Intense-dose statin, often in combination with ezetimibe	I/C	<2.6 mmol/L, or <1.8 mmol/L in presence of CVD	IIa/C
ASC If the goal is not reached with the highest tolerable statin dose, ad-on ezetimibe in post-ACS patients	IIa/B	<1.8 mmol/L, or a reduction of at least 50% if the baseline[3] LDL-C is 1.8–3.5 mmol/L	
CKD stages 3–5 are high or very high CV risk patients The use of statins or ezetimibe ad-on statin is indicated in non-dialysis dependent patients	I/A	Depends of the risk profile of the individual patient	I/A

ACS, acute coronary syndrome; FH, familial hypercholesterolemia; CKD, chronic kidney disease; LDL-C, low-density lipoprotein-cholesterol.
[1]*COR, class of recommendation.*
[2]*LOE, level of evidence.*
[3]*"baseline LDL-C" refers to the level in a subject not taking any lipid lowering medication.*

Table 3.
Recommendations for ezetimibe add-on statin combination therapy according to the 2016 ESC/EAS guidelines for the management of dyslipidemias.

with fenofibrates; in homozygote familiar hyperlipidemia (HoFH) in combination with atorvastatin or simvastatin; and in homozygous sitosterolemia (phytosterolemia) [3, 26, 35].

In clinical studies, ezetimibe, as monotherapy, reduces LDL-C in hyper-cholesterolemic patients by 15–22%. Combined therapy with statins provides an incremental reduction in LDL-C levels of 15–20%, leading to a total LDL-C reduction by 34–61%, as previously mentioned [3].

The most comprehensive data analysis for LDL-C lowering efficiency was performed by the group of Descamps, published in 2015. 27 differently designed trials (double-blind placebo and/or active controlled studies), in which statins (type of statin, statin brand, potency or dose difference) were compared with ezetimibe ad-on statin, were included, with over 21,671 patients, analyzing variables such as variances (standard deviation [SD], coefficient of variation [CV], and root mean squared error [RMSE] adjusted for various factors) for % change from baseline in LDL-C. In this very comprehensive data analysis, ezetimibe ad-on statin was found to lead to significantly more pronounced LDL-C lowering, as compared to statin monotherapy [36].

Data from a large retrospective observational study (more than 27,000 patients), published in 2014 by Toth, demonstrated a more pronounced LDL-C lowering effect of ezetimibe ad-on statin therapy, and a higher percentage of goal attainment (with respect to the risk profile of the patients), with one third of the patients not being able to attain the recommended LDL-C goal of <1.8 mmol/L. However, it was realized that there is a low prescription frequency of high-dose statins. Half of the patients (50.9%) remained on the same statin monotherapy, irrespective of their treatment goal achievement [37]. The significance of this study is even bigger given that it is a real life situation, and not a randomized study with strictly predefined inclusion criteria, study population etc.

The IMPROVE-IT study showed an on-trial average LDL-C level of 1.4 mmol/L in the simvastatin-ezetimibe group, as compared to 1.8 mmol/L in the simvastatin-monotherapy group ($p < 0.001$), leading to a total amount of LDL-C reduction of about 24% [38]. There are also a lot of small-scale studies that demonstrate superiority of ezetimibe ad-on statin therapy in terms of LDL-C lowering. For example, the Japanese study by Uemura, performed on 39 patients, compared two regimens: 10 mg atorvastatin + 10 mg ezetimibe vs. 20 mg atorvastatin in high risk patients with CAD and type 2 diabetes (T2DM). A significant improvement of the lipid profile was found in both groups in terms of total, LDL-C and high-density lipoprotein cholesterol (HDL-C), with a more pronounced improvement in the ezetimibe ad-on atorvastatin group ($p = 0.005$). A significant effect on the Apo B/Apo A-I ratio and remnant-like particle cholesterol was observed only in the atorvastatin ad-on ezetimibe treatment group. Probably the finding that gives as the most powerful information is the effect on oxidized LDL-C [malondialdehyde-modified LDL (MDA-LDL)], a form that is responsible for the proaterogenic effects of LDL-C, that was significantly more pronounced with the atorvastatin ad-on ezetimibe treatment ($p = 0.0006$) [39]. The existence of pleotropic effects, other than the hypo-lipemic effect that is widely recognized for statins, is evidentially true for ezetimibe as well. Evidence of anti-inflammatory and anti-oxidative effects is cumulating. Another Japanese study, by Tobaru, was performed on 35 CAD patients pre-treated with statins who remained above targeted LDL-C level. In terms of hypo-lipemic effect, significant additional decrease of total C, LDL-C, remnant lipoprotein C, LDL/HDL-C ratio was observed, and the percentage of patients who achieved target LDL-C level increased to 65.4% ($p = 0.001$) in the ezetimibe ad-on statin group. Although no significant effect was achieved on high-sensitive C-reactive protein (hsCRP) and oxidative stress markers, a significant reduction of tumor necrosis factor-α (TNF-α), 1.36 vs. 0.96 ($p = 0.042$) was observed [40]. On the other hand, given the IMPROVE-IT trial in which two laboratory targets were set: LDL-C (<1.8 mmol/L) and hsCRP (<2 mg/L), Bohula and colleagues summarized that ezetimibe ad-on statin treatment was far more successful in achieving both targets, or it was concluded: "Significantly more patients treated with ezetimibe/simvastatin met prespecified dual LDL-C and hsCRP targets, than patients treated with simvastatin alone (50% vs. 29%, $p < 0.001$)". Reaching both LDL-C and hsCRP targets was associated with improved outcomes after multivariable adjustment (38.9% vs. 28.0%, adjusted hazard ratio, 0.73, 0.66–0.81; $p < 0.001$) [38].

3.1.2 Effects of ezetimibe ad-on statins on ASCVD outcome

The potential benefits of adding an additional lipid lowering agent—ezetimibe on statin therapy for CVD prevention and risk reduction have been confirmed in several clinical trials.

The impact of dual lipid-lowering strategy with ezetimibe and atorvastatin on coronary plaque regression in patients with percutaneous coronary intervention: The Multicenter Randomized Controlled PRECISE-IVUS trial evaluated the effects of ezetimibe ad-on atorvastatin vs. atorvastatin monotherapy on the lipid profile and coronary atherosclerosis in Japanese patients who underwent percutaneous coronary intervention (PCI). The combination therapy resulted in lower levels of LDL-C, compared to atorvastatin monotherapy (1.6 mmol/L vs. 1.9 mmol/L; $p < 0.001$), and in the same time coronary plaque regression was observed in significantly higher percentage of patients who received atorvastatin ad-on ezetimibe (78% vs. 58%; $p = 0.004$) [41].

The majority of studies addressing the efficacy of ezetimibe ad-on statin treatment are with simvastatin, including the Simvastatin and Ezetimibe in Aortic Stenosis (SEAS) study, in patients with aortic stenosis, and the Study of Heart and Renal Protection (SHARP) (Simvastatin plus ezetimibe) trial, including 23% high risk patients with diabetes and chronic kidney disease (CKD) with or without requiring dialysis. The combination therapy demonstrated superiority over statin monotherapy in LDL-C reduction, translated in reduced primary end-point of first major ASCVD event: nonfatal MI or CV death, non-hemorrhagic stroke, or any arterial revascularization procedure, over a median follow up of 4.9 years [3, 42].

The landmark trial on ezetimibe-statin combination therapy, the largest and the longest one with ezetimibe, is the Improved Reduction of Outcomes: Vytorin Efficacy International Trial (IMPROVE-IT). A total of 18,144 patients with acute coronary syndrome (ACS) were randomized to ezetimibe (10 mg) or placebo, all receiving 40 mg simvastatin, which was increased to 80 mg if LDL-C on treatment was >2.04 mmol/L. The event rates for the primary end point at 7 years were 32.7% in the simvastatin-ezetimibe group and 34.7% in the simvastatin-monotherapy group, with an absolute risk reduction of 2% (HR 0.936; 95% CI 0.89–0.99; $p = 0.016$). Ischemic stroke was reduced by 21% ($p < 0.008$). Nevertheless, no benefit in reducing all-cause mortality or deaths from CV causes was observed, which was not unexpected, as prior trials of intensive vs. standard-dose statin therapy did not demonstrate a benefit in terms of mortality as well. There was no evidence of harm caused by the further LDL-C reduction. In this group of patients, already treated with statins to reach the goal, the absolute benefit from the added ezetimibe was small, although significant. However, the study supports the proposition that LDL-C lowering by means other than statins is beneficial and can be performed without adverse effects [38].

The diabetic sub-group analysis in the IMPROVE-IT trial provided the outcomes in 4933 (27%) patients with diabetes, one of the pre-specified trial subgroups. In this patient subset, ezetimibe ad-on statin decreased LDL-C at 1 year by 1.1 mmol/L, as compared to 0.6 mmol/L with statin monotherapy. Diabetic patients on ezetimibe ad-on statin therapy had a 14% relative risk reduction, or 5.5% absolute reduction, compared with a 2% absolute risk reduction for non-diabetics. The most notorious reductions were seen regarding ischemic stroke (39%), MI (24%), and the composite of death due to CV causes, MI or stroke (20%). These CV effects of ezetimibe ad-on statin therapy are considered to be a result of the more prominent reduction of LDL-C (mean 0.5 mmol/L), compared to simvastatin monotherapy, with an average value of 1.4 mmol/L. This sub-study analysis demonstrated superiority of the statin-ezetimibe combination therapy in CV prevention in diabetic subsets especially [26, 34, 35, 38].

Another significant effect of ezetimibe ad-on statin therapy is cerebrovascular protection. The 2013 ACC/AHA cholesterol guideline recommends the use of ezetimibe as an ad-on statin, additional LDL-C lowering agent in stroke patients [35]. The advantage of ezetimibe ad-on statin therapy in this patient subgroup was observed in the IMPROVE-IT study. The highest risk benefit was observed in the subgroup of patients with ischemic CVD with a 21% relative reduction of ischemic stroke ($p < 0.008$). The addition of ezetimibe as a non-statin type drug, to statin treatment contributed to further reduction of LDL-C, which translated into additional decrease in reoccurrence and mortality of/from cerebrovascular events. Achieving target values with ezetimibe ad-on statin combination allows administration of low to moderate-dose statin, which decreases the risks of adverse effects related to high-dose statin therapy [43].

The current trial results make it obvious that the higher the risk profile of the patient is, the bigger is the benefit, in terms of risk reduction, when ezetimibe is ad-on statin treatment. Taken together, all these studies support the decision to propose ezetimibe as a second-line therapy, in association with statins, when the therapeutic goal is not achieved with the maximal tolerated statin dose or in patients intolerant or with contraindications to these drugs [3, 35].

3.2 Combination therapy: PCSK9 inhibitors ad-on statin

3.2.1 Key points

- The FDA approved the first PCSK9 inhibitor in 2015
- There are currently two PCSK9 inhibitors on the market, alirocumab and evolocumab
- There was a third PCSK9 inhibitor—bococizumab, but its' development was discontinued by Pfizer in late 2016. The key reasons for this were a high level of immunogenicity and wide variability in the LDL-C lowering response. *Immunogenicity*: in statin-treated patients, PCSK9 inhibition with bococizumab led to a short-term LDL-C reduction of 55–60%. However, this effect was attenuated over time in 10–15% of patients due to the development of anti-drug antibodies. This effect was specific to bococizumab, which is a partially humanized monoclonal antibody, characterized by substitution of rodent deoxyribonucleic acid (DNA) sequences for <5% of human DNA sequences. It is thought that this substitution may have directly affected the immunogenicity of the antibody. This effect has not been reported for either evolocumab or alirocumab, which are fully human monoclonal antibodies. This immunogenicity may also explain the higher rate of injection site reactions (~10%) observed with bococizumab, compared with either alirocumab or evolocumab (<5%). *Variability in LDL-C lowering response*: Irrespective of the presence or absence of antidrug antibodies, there was wide individual variability in the LDL-C lowering response with bococizumab; about 1 in 10 showed no reduction in LDL-C levels
- Patients with familial hyperlipidemia and those with clinical ASCVD, not reaching lipid-reducing goals, including those with statin intolerance, are at greatest need of PCSK9 inhibitors, because no adequate alternative treatment exists
- Multiple guidelines with different approaches to lipid treatment have created confusion among clinicians; thus, defining the patients with ASCVD, or at high CV risk, who have not met LDL-C treatment goals is complicated
- Although PCSK9 inhibitors seem to support the LDL-C hypothesis (the lower the LDL-C level, the lower the CV risk), results of ongoing long-term outcome studies are yet to be presented
- Prescribing PCSK9 inhibitors will likely be limited by economics rather than by clinicians' judgment about the best interest of their patients [44].

Many believe that proprotein convertase subtilisin/kexin type 9 (PCSK9) inhibitors are the pharmacotherapeutic innovation of the past 2 decades in terms of CV events prevention.

The 2017 Update of ESC/EAS Task Force on practical clinical guidance for proprotein convertase subtilisin/kexin type 9 inhibition in patients with atherosclerotic cardiovascular disease or in familial hypercholesterolemia defined:

Patients with clinical ASCVD and substantially elevated LDL-C levels (patients already on maximally tolerated statin therapy (ideally with concomitant ezetimibe), or unable to tolerate three or more statins), and,

Familial hypercholesterolemia (FH) patients without clinical ASCVD but with substantially elevated LDL-C levels (patients on maximally tolerated statin therapy plus ezetimibe), as priority patient groups for PCSK9 inhibitors (**Figure 1**) [45].

3.2.2 LDL-C lowering and PCSK9 inhibitors: what have we learned?

The Further Cardiovascular Outcomes Research With PCSK9 Inhibition in Subjects With Elevated Risk (FOURIER) trial, aimed to evaluate the efficacy of evolocumab, compared to placebo, in patients with clinically evident CVD. 69% of the patients were on a high-intensity statin, while 30% were on a moderate-intensity statin therapy, randomized to evolocumab 140 mg subcutaneous every 2 weeks or 420 mg monthly (n = 13,784) vs. placebo every 2 weeks (n = 13,780). Evolocumab led to a 59% LDL-C level reduction (from 2.4 mmol/L to 0.78 mmol/ L), with an absolute LDL-C reduction of 1.4 mmol/L [46].

The Goal Achievement After Utilizing an Anti-PCSK9 Antibody in Statin Intolerant Subjects-3 (GAUSS-3) trial aimed to evaluate the effect of 24 weeks

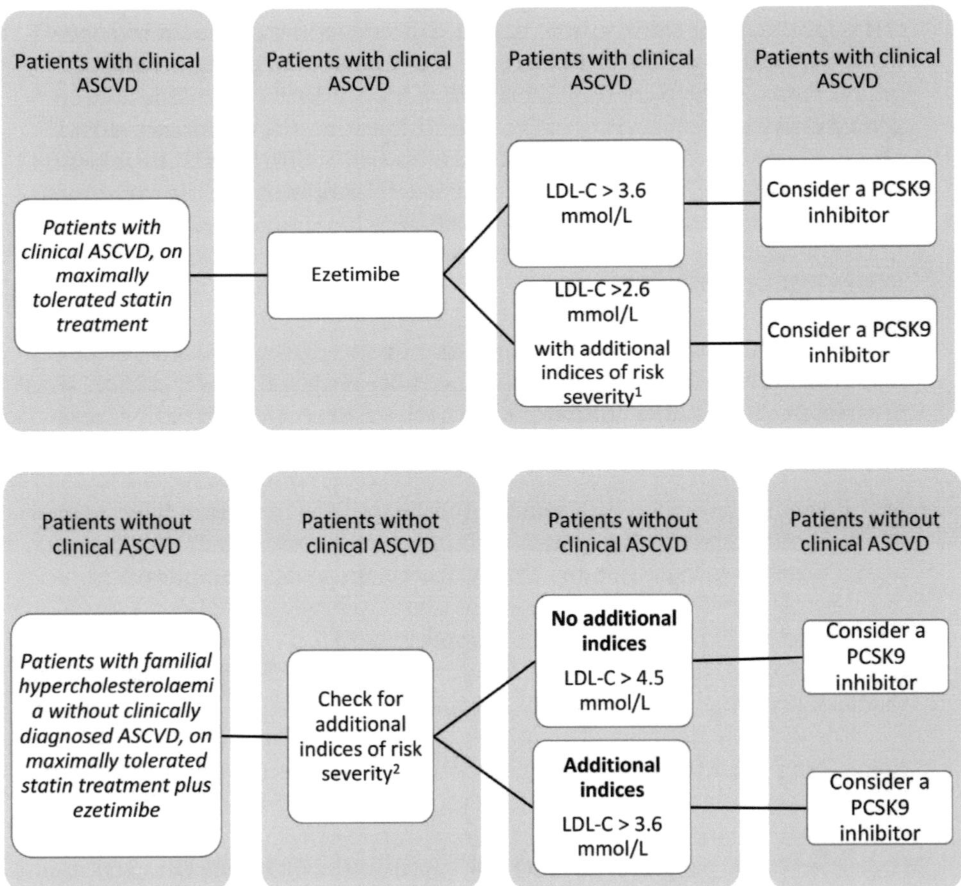

Figure 1.
Appropriate use of PCSK9 inhibitor, as recommended in 2017 update of ESC/ESC task force on practical clinical guidance for proprotein convertase subtilisin/kexin type 9 inhibition in patients with atherosclerotic cardiovascular disease or in familial hypercholesterolemia. ASCVD, atherosclerotic cardiovascular disease; LDL-C, low-density lipoprotein-cholesterol; [1]Including: familial hypercholesterolemia; diabetes mellitus with target organ damage, or a major risk factor; severe and/or extensive ASCVD; rapid progression of ASCVD (i.e. repeated ACS, unplanned coronary revascularizations). [2]Including: diabetes mellitus with target organ damage, or a major risk factor; Lipoprotein(a) > 50 mg/dL; major risk factors: smoking, marked hypertension; >40 years of age without treatment; premature ASCVD (<55 years in males and <60 years in females) in first-degree relatives; imaging indicators.

of evolocumab administered subcutaneously (SC) every month, compared with ezetimibe, on LDL-C levels in adults with high cholesterol, who are unable to tolerate an effective dose of a statin due to muscle-related side effects (MRSE). Evolocumab produced significantly larger reductions in LDL-C levels, compared to ezetimibe (16.7% reduction with ezetimibe and a more than 50% reduction with evolocumab). Despite very high baseline values, the LDL-C goal of less than 1.8 mmol/L was achieved in nearly 30% of evolocumab-treated patients and 1.4% of ezetimibe-treated patients. The LDL-C reduction for both drugs was stable for 4 weeks and sustained during the course of the 24 weeks of treatment [47].

The effect of evolocumab or ezetimibe added to moderate- or high-intensity statin therapy on LDL-C lowering in patients with hypercholesterolemia—the LAPLACE-2 randomized clinical trial evaluated the efficacy and tolerability of evolocumab when used in combination with a moderate- vs. high-intensity statin. 2067 patients with primary hypercholesterolemia and mixed dyslipidemia were randomized to 24 treatment groups. Patients were initially randomized to a daily, moderate-intensity (atorvastatin 10 mg, simvastatin 40 mg, or rosuvastatin 5 mg) or high-intensity (atorvastatin 80 mg, or rosuvastatin 40 mg) statin. After a 4-week lipid-stabilization period, patients were randomized to compare evolocumab (140 mg every 2 weeks) vs. placebo (every 2 weeks) or ezetimibe (10 mg daily; atorvastatin patients only) when added to statin therapies. In patients treated with atorvastatin (10 mg or 80 mg), the addition of ezetimibe resulted in LDL-C reductions by 17–24% from baseline, compared with the addition of evolocumab, administered every 2 weeks, which reduced LDL-C values by 61–62% (treatment differences vs. placebo and ezetimibe both significant [$p < 0.001$]. For patients receiving a moderate-intensity statin, evolocumab reduced LDL-C values from a baseline mean of 3.1 mmol/L to an on-treatment mean of 1.2 mmol/L, and 88–94% of the patients achieved target LDL-C levels, less than 1.8 mmol/L. For patients receiving a high-intensity statin, evolocumab reduced LDL-C values from a baseline mean of 2.4 mmol/L to an on-treatment mean of 0.9 mmol/L, and 94% achieved the target LDL-C value. In the atorvastatin-treated patients, addition of ezetimibe resulted in achievement of an LDL-C level less than 1.8 mmol/L in 17–20% of patients receiving moderate-intensity statins and 51–62% of those receiving high-intensity statins, vs. 86–94% of patients achieving target LDL-C values in the evolocumab-atorvastatin group [31].

The efficacy and safety of alirocumab in high cardiovascular risk patients with inadequately controlled hypercholesterolemia on maximally tolerated doses of statins: the ODYSSEY COMBO II randomized controlled trial aimed to compare efficacy, in terms of LDL-C lowering, and safety of alirocumab vs. ezetimibe as ad-on therapy to maximally tolerated statin treatment in high CV risk patients with inadequately controlled hypercholesterolemia. Patients were randomized to subcutaneous alirocumab 75 mg every 2 weeks (plus oral placebo) or oral ezetimibe 10 mg daily (plus subcutaneous placebo) on a background of statin therapy. At week 24, mean ± SE reductions in LDL-C from baseline were 50.6 ± 1.4% for alirocumab vs. 20.7 ± 1.9% for ezetimibe ($p < 0.0001$). 77.0% of alirocumab and 45.6% of ezetimibe patients achieved target LDL-C values of <1.8 mmol/L ($p < 0.0001$). Mean achieved LDL-C levels, by week 24, were 1.3 mmol/L with alirocumab and 2.1 mmol/L with ezetimibe [32].

3.2.3 Cardiovascular outcome studies with PCSK9 inhibitors: what have we learned?

The FOURIER trial, evaluating the effect of evolocumab on the risk of CV death, MI, stroke, hospitalization for unstable angina, or coronary revascularization, in 27,564 patients with clinically evident CVD, is the first randomized study to be completed, regarding PCSK9 inhibitors long-term efficacy and safety. The primary outcome, incidence of CV death, MI, stroke, hospitalization for unstable angina, or coronary revascularization, occurred in 12.6% of the evolocumab group vs. 14.6% of the placebo group ($p < 0.0001$). There was a 9.8% absolute MACE rate reduction, compared to 11.3% with placebo, over 2.2 years, with a relative risk reduction of 15%. This finding was consistent among all tested subgroups. Benefit was enhanced among higher-risk subgroups (those with recent MI, multiple prior MIs, and residual multivessel coronary artery disease), compared to those without such characteristics. There was a linear relationship between LDL-C and adverse CV events, such that adverse events continued to decline to the lowest levels of LDL-C ($p = 0.0012$). Among those with baseline LDL-C < 1.8 mmol/L, evolocumab reduced the primary endpoint (hazard ratio [HR] 0.80, 95% confidence interval [CI] 0.60–1.07) to a similar degree as those with baseline LDL-C ≥ 1.8 mmol/L (HR 0.86, 95% CI 0.79–0.92; $p = 0.65$ for interaction).

There was a greater absolute reduction in major adverse events for evolocumab vs. placebo among those with the highest baseline inflammatory risk (among those with high-sensitivity C-reactive protein <1 mg/dl, 1–3 mg/dl, and >3 mg/dl, there was an absolute reduction in the primary outcome of 1.6, 1.8, and 2.7%, respectively). PCSK9 inhibition represents a novel approach to lower LDL-C levels and improves cardiovascular outcomes. However, for the duration of follow-up, there was no benefit on cardiovascular or all-cause mortality, and cost remains an issue [46].

Regarding alirocumab, cardiovascular outcomes and safety will be assessed in an ongoing study, the ODYSSEY Outcomes: Evaluation of Cardiovascular Outcomes After an Acute Coronary Syndrome During Treatment With Alirocumab trial. 18,600 patients, who have experienced an acute coronary syndrome (ACS), are allocated to alirocumab or placebo, for a maximum duration of 64 months. The primary objective of the trial is to compare the effect of alirocumab with placebo on the occurrence of cardiovascular events (composite endpoint of coronary heart disease (CHD) death, non-fatal MI, fatal and non-fatal IS, unstable angina requiring hospitalization). No results are reported yet [32].

4. Safety profiles of statin monotherapy and statin combination therapy

4.1 Adverse effects of statin therapy

The only excesses of adverse events that have been reliably demonstrated to be caused by statin therapy are myopathy and diabetes mellitus, along with a probable excess of hemorrhagic stroke. However, the absolute risks of these adverse effects remain small, by comparison with the absolute benefits [25].

4.1.1 Increases in myopathy rates

Approved statin regimens have been associated, both in observational studies and in randomized trials, with large relative risks for myopathy, but typically with small absolute excesses (about 1 case per 10,000 people treated per year), and even smaller excesses in the incidence of rhabdomyolysis (about 2–3 cases per 100,000 treated per year). It usually resolves rapidly when statin therapy is . [48]. The risk of myopathy is dose related and it appears to depend on the statin circulation levels. In the Study of the Effectiveness of Additional Reductions in Cholesterol and Homocysteine (SEARCH) randomized trial, simvastatin 80 mg daily produced a more than ten-fold higher rate (at least 1 case of myopathy per 1000 patients treated yearly), compared to 20 mg daily (about 1 case per 10,000 yearly), so the high-dose regimen is no longer recommended routinely [49].

4.1.2 Increases in diabetes rates

In the Justification for the Use of Statins in Primary Prevention: An Intervention Trial Evaluating Rosuvastatin (JUPITER) randomized trial, among 17,802 patients without a history of vascular disease, concentrations of glycated hemoglobin, after 2 years, were slightly higher among the patients allocated to rosuvastatin 20 mg daily compared to those allocated to placebo (5.9 vs. 5.8%; p = 0.001). There was also a small excess of newly diagnosed diabetes (3.0 vs. 2.4%; p = 0.01), which corresponds to a 25% proportional increase. In subsequent meta-analyses, standard statin dose regimens were associated with a proportional increase of about 10% in reported diabetes, and more intensive statin regimens (as used in JUPITER) with about a 10% further increase. This excess of diabetes diagnoses appeared soon after the start of statin therapy, mainly among patients who had previous risk factors for diabetes [50].

4.1.3 Probable increases in hemorrhagic stroke rates

In The Stroke Prevention by Aggressive Reduction of Cholesterol Levels (SPARCL) trial, among 4731 patients with prior cerebrovascular disease, allocation to atorvastatin 80 mg daily produced a definite reduction in ischemic stroke (218 [9.2%] vs. 274 [11.6%]; p = 0.008), but there was also a possible increase in hemorrhagic stroke (55 [2.3%] vs. 33 [1.4%]; p = 0.02). When these results were combined with those from the other trials included in the CTT meta-analysis, there was a 21% (95% CI 5–41; p = 0.01) proportional increase in the incidence of hemorrhagic stroke per mmol/L reduction in LDL-C [51].

4.2 Safety profile of ezetimibe ad-on statins combination therapy

The relationship between lipid-lowering medications, glycemic control, insulin resistance and new-onset diabetes has been studied since the introduction of hipo-lipemic medications. We know that glycemic control is impaired not only by statin treatment but also with niacin. At the opposite, bile-acid sequestrates demonstrate moderate lipid and glucose lowering effects, and fibrates (particularly bezafibrate) may produce beneficial effects on glucose metabolism and insulin sensitivity. Considering statins, as the most widely used hypo-lipemic drugs, this is an important issue. Statins lead to a mild elevation of hemoglobin A1c (HgbA1c) and fasting plasma glycose (FPG), and increase the incidence of new-onset diabetes, an effect known to be dose and agent dependent (Pravastatin and Pitavastatin have less diabetogenic effect and positive impact on insulin sensitivity). The aforementioned is most pronounced in patients with baseline impaired fasting blood glucose (FBG), at older age and with metabolic syndrome. However, it has been demonstrated that the risk of new-onset diabetes is overweight by the benefit of CV risk reduction [3].

For a long period of time there was a lack of clinical trials addressing the same question in ezetimibe treatment, but data was gathered from experimental animal studies that described how ezetimibe ameliorates metabolic markers, such as hepatic steatosis and insulin resistance. The process is via inhibition of the intestinal cholesterol absorption, and inhibition of the hepatic NPC1L1, leading to decreased hepatic insulin resistance, improved glycemic control and insulin sensitivity, especially in patients with metabolic disorders (obesity and hepatic steatosis). This was harder to prove in humans, as ezetimibe is usually used as statin co-therapy and individual impact of ezetimibe cannot be evaluated. In a recently published systematic review of randomized clinical trials, performed by Wu and co-authors on 2440 patients, experimental data was confirmed. Ezetimibe did not cause any adverse effects in terms of increased levels of FBG and HbA1c. Compared with high-dose statin therapy, ezetimibe ad-on low-dose statin for more than 3 months may even have beneficial effects on glycemic control [52].

Statin associated muscle symptoms are a very common side effect, also known to be dose-dependent. It seems that ezetimibe ad-on low dose statin therapy is one of the possibilities to achieve good LDL-C control and CV risk reduction with lesser side effects, as demonstrated with myalgia [53]. The 2016 ESC/EAS Guidelines for the management of dyslipidemias recommend ezetimibe to be considered in combination with a low-dose statin or second- or third-line statin in order to manage statin-attributed muscle symptoms [3].

4.3 Safety profile of PCSK9 inhibitors ad-on statins combination therapy

Despite this new evidence from the FOURIER trial, gaps remain in our knowledge regarding the use of PCSK9 inhibition in clinical practice. The ODYSSEY Outcomes trial will provide additional information in patients treated with a PCSK9 inhibitor within 1–12 months [45].

As with all novel treatments, long-term safety remains to be established. To date there are exposure data for up to 4 years' treatment with a PCSK9 inhibitor, involving a background of concomitant statin therapy. Potential injection site reactions occurred in <5% of patients, and were mainly of very mild intensity with no evidence of a cumulative effect. When the PCSK9 inhibitor was compared to the previous standard of care (statin with or without ezetimibe), annualized event rates for muscle symptoms (4.7% vs. 8.5% with standard of care), and new-onset diabetes (2.8% vs. 4.0%, respectively) appeared similar [45].

The safety of very low LDL-C levels merits special consideration, given that one in four patients treated with evolocumab in FOURIER attained LDL-C levels less than 0.5 mmol/L. Evidence to date suggests no detrimental impact on steroid hormone production, enterohepatic circulation of bile acids, or neuronal cell function. Indeed, these LDL-C levels are also consistent with the very low levels observed in newborns which, despite the physiological and developmental demands of infancy, are compatible with normal development [54].

Additionally, data from the ODYSSEY program, the FOURIER and 6-year follow-up from the IMPROVE-IT trial showed no increase in adverse events including severe muscle symptoms, liver enzyme elevation, cognitive adverse events, or hemorrhagic stroke with very low LDL-C levels [45].

5. The war of today

We now know the battle is going to be very hard. The "old ones" are not ready to go to history, while the "young ones" are still to be proven. What does the newest published data say? In January 2018 Khan and co-authors published a meta-analysis of statins, PCSK9 inhibitors and ezetimibe, the later two with or without statins, regarding ASCVD reduction benefit. The most comprehensive meta-analysis included 189,116 patients from 39 randomized control trials. PCSK9 inhibitors were ranked as the best treatment for prevention of major adverse cardiovascular events: myocardial infarction and stroke. Statins were ranked as the most effective therapy for reducing all-cause and CV mortality. In terms of reduction of CV mortality PCSK9 inhibitors were ranked as the second best treatment followed by ezetimibe ad-on statin [55].

6. Conclusion

Statins remain the cornerstone anti-lipemic treatment, proven to be very effective in reducing ASCVD risk. Ezetimibe ad-on statin combination therapy is an effective treatment that leads to additional LDL-C lowering, recommended in situations where with maximal or maximally tolerated statin monotherapy treatment LDL-C target goals cannot be achieved. It leads to an additional CVD risk reduction, and in the same time is safe, with a possible beneficial effect over the statin adverse influence on glycemic metabolism. Having in mind the evidence from the first of the cardiovascular outcomes studies with PCSK9 inhibitors, the addition of a PCSK9 inhibitor should be considered in patients with ASCVD, and in FH patients without a prior clinical event, who have substantially elevated LDL-C levels despite maximally tolerated statin with or without ezetimibe therapy, or inability to tolerate appropriate doses of at least three statins. Prioritizing the use of combination therapy in these specific patient groups may help reduce cardiovascular outcomes and the impact of the associated physical and/or psychological disability.

Acknowledgements

Both authors equally contributed to this chapter.

Conflict of interest

We declare no conflict of interest. This chapter was written independently. No company or institution supported it financially.

Abbreviations

AACE	American Association of Clinical Endocrinologists
ABC	ATP-binding cassette
ACC/AHA	American College of Cardiology/American Heart Association
ACE	American College of Endocrinology
ACS	acute coronary syndrome
ApoB	apolipoprotein B
ARIC study	the Atherosclerosis Risk in Communities study
ASCVD	atherosclerotic cardiovascular disease
CAD	coronary artery disease
CHD	coronary heart disease
CKD	chronic kidney disease
CTT	Cholesterol Treatment Trialists' Collaboration
CV	cardiovascular
CV	coefficient of variation
CVD	cardiovascular disease
CYP450	cytochrome P450
DNA	deoxyribonucleic acid
ESC/EAS	European Society of Cardiology/European Atherosclerosis Society
FBG	fasting blood glucose
FDA	Food and Drug Administration
FH	familial hypercholesterolemia

FOURIER trial	Further Cardiovascular Outcomes Research With PCSK9 Inhibition in Subjects With Elevated Risk trial
FPG	fasting plasma glycose
GAUSS-3 trial	the Goal Achievement After Utilizing an Anti-PCSK9 Antibody in Statin Intolerant Subjects-3 trial
HDL-C	high-density lipoprotein cholesterol
HgBA1C	hemoglobin A1c
HMG-CoA	3-hydroxy-3-methylglutaryl coenzyme A
HoFH	homozygote familiar hyperlipidemia
hsCRP	high-sensitive C-reactive protein
IDEAL trial	the Incremental Decrease in Endpoints Through Aggressive Lipid Lowering trial
IDL	intermediate density lipoproteins
IgG1	immunoglobulin G1
IMPROVE-IT study	Examining Outcomes in Subjects With Acute Coronary Syndrome: Vytorin vs. Simvastatin
IS	ischemic stroke
JUPITER trial	in the Justification for the Use of Statins in Primary Prevention: An Intervention Trial Evaluating Rosuvastatin randomized trial
LAPLACE-2 trial	the LDL-C Assessment With Proprotein Convertase Subtilisin Kexin Type 9 Monoclonal Antibody Inhibition Combined With Statin Therapy 2 trial
LDL	low-density lipoprotein
LDL-C	low-density lipoprotein cholesterol
LDL-R	LDL receptor
Lp(a)	lipoprotein(a)
MDA-LDL	malondialdehyde-modified LDL
MI	myocardial infarction
MoAbs	monoclonal antibodies
MRSE	muscle-related side effects
MVE	major vascular events
NPC1L1	Niemann Pick C1 like 1 protein
ODDYSSEY trial	the Reductions in Atherogenic Lipids and Major Cardiovascular Events: A Pooled Analysis of 10 ODYSSEY Trials Comparing Alirocumab With Control
PCI	percutaneous coronary intervention
PCSK9	Proprotein Convertase Subtilisin/Kexin Type 9
PRECISE-IVUS trial	the Impact of Dual Lipid-Lowering Strategy With Ezetimibe and Atorvastatin on Coronary Plaque Regression in Patients With Percutaneous Coronary Intervention: The Multicenter Randomized Controlled trial
PROVE-IT trial	Pravastatin or Atorvastatin Evaluation and Infection Therapy trial
RCTs	randomized clinical trials
RMSE	root mean squared error
SC	subcutaneously
SD	standard deviation
SEARCH trial	in the Study of the Effectiveness of Additional Reductions in Cholesterol and Homocysteine randomized trial
SEASE study	Simvastatin and Ezetimibe in Aortic Stenosis study
SHARP trial	the Study of Heart and Renal Protection (Simvastatin plus ezetimibe) trial

SLC	solute carrier
SPARCL trial	in The Stroke Prevention by Aggressive Reduction of Cholesterol Levels trial
T2DM	type 2 diabetes mellitus
TNF-α	tumor necrosis factor-α
TNT trial	treating to New Targets trial
UGT	UDP-glucuronosyltransferase
VLDL	very low-density lipoproteins

Author details

Marija Vavlukis[1] and Ana Vavlukis[2]

1 University Clinic of Cardiology, Medical Faculty, Ss' Cyril and Methodius University, Skopje, Republic of Macedonia

2 Faculty of Pharmacy, Ss' Cyril and Methodius University, Skopje, Republic of Macedonia

*Address all correspondence to: marija.vavlukis@gmail.com

References

[1] Ference BA, Ginsberg HN, Graham I, et al. Low-density lipoproteins cause atherosclerotic cardiovascular disease. 1. Evidence from genetic, epidemiologic, and clinical studies. A consensus statement from the European Atherosclerosis Society Consensus Panel. European Heart Journal. 2017;**38**(32):2459-2472. DOI: 10.1093/eurheartj/ehx144

[2] Nozue T. Lipid lowering therapy and circulating PCSK9 concentration. Journal of Atherosclerosis and Thrombosis. 2017;**24**(9):895-907. DOI: 10.5551/jat.RV17012

[3] Catapano AL, Graham I, De Backer G, et al. The Task Force for the Management of Dyslipidaemias of the European Society of Cardiology (ESC) and European Atherosclerosis Society (EAS) developed with the special contribution of the European Assocciation for Cardiovascular Prevention & Rehabilitation (EACPR). 2016 ESC/EAS Guidelines for the Management of Dyslipidaemias. European Heart Journal. 2016;**37**:2999-3058. DOI: 10.1093/eurheartj/ehw272

[4] Robinson JG. Nonstatins and Proprotein Convertase Subtilisin/Kexin Type 9 (PCSK9) inhibitors: Role in non-familial hypercholesterolemia. Progress in Cardiovascular Diseases. 2016;**59**(2):165-171. DOI: 10.1016/j.pcad.2016.07.009

[5] Ramkumar S, Raghunath A, Raghunath S. Statin therapy: Review of safety and potential side effects. Acta Cardiologica Sinica. 2016;**32**(6):631-639. DOI: 10.6515/ACS20160611A

[6] McFarland AJ, Anoopkumar-Dukie S, Arora DS, et al. Molecular mechanisms underlying the effects of statins in the central nervous system. Harry GJ, ed. International Journal of Molecular Sciences. 2014;**15**(11):20607-20637. DOI: 10.3390/ijms151120607

[7] Gazzerro P, Proto MC, Gangemi G, et al. Pharmacological actions of statins: A critical appraisal in the management of cancer. Pharmacological Reviews. 2012;**64**(1):102-146. DOI: 10.1124/pr.111.004994

[8] Meor Anuar Shuhaili MFR, Samsudin IN, Stanslas J, et al. Effects of different types of statins on lipid profile: A perspective on Asians. International Journal of Endocrinology and Metabolism. 2017;**15**(2):e43319. DOI: 10.5812/ijem.43319

[9] Smith BA, Wright C, Davidson M. Role of Ezetimibe in lipid-lowering and cardiovascular disease prevention. Current Atherosclerosis Reports. 2015;**17**:72. DOI: 10.1007/s11883-015-0550-5

[10] Hammersley D, Signy M. Ezetimibe: An update on its clinical usefulness in specific patient groups. Therapeutic Advances in Chronic Disease. 2017;**8**(1):4-11. DOI: 10.1177/2040622316672544

[11] Tie C, Gao K, Zhang N, et al. Ezetimibe attenuates atherosclerosis associated with lipid reduction and inflammation inhibition. PLoS One. 2015;**10**(11):e0142430. DOI: 10.1371/journal.pone.0142430

[12] Kim H, Choi HY, Kim YH, et al. Pharmacokinetic interactions and tolerability of rosuvastatin and ezetimibe: An open-label, randomized, multiple-dose, crossover study in healthy male volunteers. Drug Design, Development and Therapy. 2018;**12**:815-821. DOI: 10.2147/DDDT.S158408

[13] Nabil GS. Proprotein Convertase Subtilisin Kexin 9 (PCSK9) inhibitors in the treatment of

hypercholesterolemia and other pathologies. Current Pharmaceutical Design. 2013;**19**:3161-3172. DOI: 10.2174/1381612811319990313

[14] Page MM, Watts GF. PCSK9 inhibitors—Mechanisms of action. Australian Prescriber. 2016;**39**(5):164-167. DOI: 10.18773/austprescr.2016.060

[15] Giugliano RP, Sabatine MS. Are PCSK9 inhibitors the next breakthrough in the cardiovascular field? Journal of the American College of Cardiology. 2015;**65**(24):2638-2651. DOI: 10.1016/j.jacc.2015.05.001

[16] Hess PL, Kennedy K, Cowherd M, et al. Implications of the FDA approval of PCSK9 inhibitors and FOURIER results for contemporary cardiovascular practice: An NCDR research to practice (R2P) project. American Heart Journal. 2018;**195**:151-152. DOI: 10.1016/j.ahj.2017.09.004

[17] Fala L. Repatha (Evolocumab): Second PCSK9 inhibitor approved by the FDA for patients with familial hypercholesterolemia. American Health & Drug Benefits. 2016;**9** (Spec Feature):136-139

[18] Manniello M, Pisano M. Alirocumab (Praluent): First in the new class of PCSK9 inhibitors. Pharmacy and Therapeutics. 2016;**41**(1):28-53

[19] Carroll MD, Kit BK, Lacher DA, Shero ST, Mussolino ME. Trends in lipids and lipoproteins in US adults, 1988-2010. Journal of the American Medical Association. 2012;**308**:1545-1554

[20] Stone NJ, Robinson JG, Lichtenstein AH, et al. 2013 ACC/AHA guideline on the treatment of blood cholesterol to reduce atherosclerotic cardiovascular risk in adults: A report of the American College of Cardiology/American Heart Association Task Force on Practice Guidelines. Circulation. 2014;**129** (suppl 2):S1-S45

[21] Soran H, Schofield JD, Durrington PN. Cholesterol, not just cardiovascular risk, is important in deciding who should receive statin treatment. European Heart Journal. 2015;**36**:2975-2983

[22] Cholesterol Treatment Trialists' (CTT) Collaboration. Efficacy and safety of more intensive lowering of LDL cholesterol: A meta-analysis of data from 170 000 participants in 26 randomised trials. Lancet. 2010;**376**:1670-1681

[23] Cholesterol Treatment Trialists' (CTT) Collaborators. The effects of lowering LDL cholesterol with statin therapy in people at low risk of vascular disease: Meta-analysis of individual data from 27 randomised trials. Lancet. 2012;**380**:581-590

[24] Soran H, Dent R, Durrington P. Evidence-based goals in LDL-C reduction. Clinical Research in Cardiology. 2017;**106**:237-248. DOI: 10.1007/s00392-016-1069-7

[25] Collins R, Reith C, Emberson J, et al. Interpretation of the evidence for the efficacy and safety of statin therapy. Lancet. 2016;**388**:2532-2561. DOI: 10.1016/ S0140-6736(16)31357-5

[26] Jellinger PS, Handelsman Y, Rosenblit PD, et al. AACE 2017 Guidelines, American Association of Clinical Endocrinologists and American College of Endocrinology Guidelines for Management of Dyslipidemia and Prevention of Cardiovascular Disease. Endocrine Practice. 2017;**23**(Suppl 2):1-63

[27] LaRosa JC, Grundy SM, Waters DD, et al. Intensive lipid lowering with atorvastatin in patients with stable coronary disease. The New England Journal of Medicine. 2005;**352**:1425-1435. DOI: 10.1056/NEJMoa050461

[28] Pedersen TR, Faergeman O, Kastelein JJ, et al. High-dose

atorvastatin vs usual-dose simvastatin for secondary prevention after myocardial infarction the IDEAL study: A randomized controlled trial. Journal of the American Medical Association. 2005;**294**(19):2437-2445. DOI: 10.1001/jama.294.19.2437

[29] Cannon CP, Braunwald E, McCabe CH, et al. Intensive versus moderate lipid lowering with statins after acute coronary syndromes. The New England Journal of Medicine. 2004;**350**:1495-1504. DOI: 10.1056/NEJMoa040583

[30] Baer J. AACE and EAS lipid guidelines. Expert analysis. AACE and EAS lipid guidelines—American College of Cardiology [internet]. http://www.acc.org/latest-in-cardiology/articles/2017/08/11/08/35/aace-and-eas-lipid-guidelines. [Accessed: 2018-10-05]

[31] Robinson JG, Rogers WJ, Nedergaard BS, et al. Rationale and design of LAPLACE-2: A phase 3, randomized, double-blind, placebo- and Ezetimibe-controlled trial evaluating the efficacy and safety of Evolocumab in subjects with hypercholesterolemia on background statin therapy. Clinical Cardiology. 2014;**37**(4):195-203. DOI: 10.1002/clc.22252

[32] Ray KK, Ginsberg HN, Davidson MH, et al. Reductions in atherogenic lipids and major cardiovascular events: A pooled analysis of 10 ODYSSEY trials comparing alirocumab with control. Circulation. 2016;**134**(24):1931-1943. DOI: 10.1161/CIRCULATIONAHA.116.024604

[33] Sabatine MS, Giugliano RP, Keech AC, et al. Evolocumab and clinical outcomes in patients with cardiovascular disease. The New England Journal of Medicine. 2017;**376**(1713):1722. DOI: 10.1056/NEJMoa1615664

[34] Cannon CP, Blazing MA, Giugliano RP, et al. Ezetimibe added to statin therapy after acute coronary syndromes. The New England Journal of Medicine. 2015;**372**(2387):2397. DOI: 10.1056/NEJMoa1410489

[35] Lloyd-Jones D, Morris P, Ballantyne C, et al. 2016 ACC expert consensus decision pathway on the role of non-statin therapies for LDL-cholesterol lowering in the management of atherosclerotic cardiovascular disease risk: A report of the American College of Cardiology Task Force on Clinical Expert Consensus Documents. Journal of the American College of Cardiology. 2016;**68**:92-125. DOI: 10.1016/j.jacc.2016.03.519

[36] Descamps O, Tomassini JE, Lin J, et al. Variability of the LDL-C lowering response to ezetimibe and ezetimibe + statin therapy in hypercholesterolemic patients. Atherosclerosis. 2015;**240**(2):482-489. DOI: 10.1016/j.atherosclerosis.2015.03.004

[37] Toth PP, Foody JM, Tomassini JE, et al. Therapeutic practice patterns related to statin potency and ezetimibe/simvastatin combination therapies in lowering LDL-C in patients with high-risk cardiovascular disease. Journal of Clinical Lipidology. 2014;**8**(1):107-116. DOI: 10.1016/j.jacl.2013.09.009

[38] Bohula E, Giugliano R, Cannon C, et al. Achievement of dual low-density lipoprotein cholesterol and high-sensitivity C-reactive protein targets more frequent with the addition of ezetimibe to simvastatin and associated with better outcomes in IMPROVE-IT. Circulation. 2015;**132**:1224-1233. DOI: 10.1161/CIRCULATIONAHA.115.018381

[39] Uemura Y, Watarai M, Ishii H, et al. Atorvastatin 10 mg plus ezetimibe 10 mg compared with atorvastatin 20 mg: Impact on the lipid profile in Japanese patients with abnormal glucose tolerance and coronary artery disease. Journal of Cardiology. 2012;**59**(1):50-56. DOI: 10.1016/j.jjcc.2011.09.001

[40] Tobaru T, Seki A, Asano R, et al. Lipid-lowering and anti-inflammatory effect of ezetimibe in hyperlipidemic patients with coronary artery disease. Heart and Vessels. 2013;**28**(1):39-45. DOI: 10.1007/s00380-012-0243-8

[41] Tsujita K, Sugiyama S, Sumida H, et al. Impact of dual lipid-lowering strategy with Ezetimibe and atorvastatin on coronary plaque regression in patients with percutaneous coronary intervention: The multicenter randomized controlled PRECISE-IVUS trial. Journal of the American College of Cardiology. 2015;**66**(5):495-507. DOI: 10.1016/j.jacc.2015.05.065

[42] Mancini J, Hegele R, Leiter L. Dyslipidemia. Canadian Diabetes Association clinical practice guidelines expert committee. Canadian Journal of Diabetes. 2013;**37**(Suppl 1):S110-S116. DOI: 10.1016/j.jcjd.2013.01.032

[43] Lovadi E, Csécsei P, Lovig C, et al. Lipids and cerebrovascular disease—New therapeutic options in lowering LDL-cholesterol. Orvosi Hetilap. 2016;**157**(52):2059-2065. DOI: 10.1556/650.2016.30643

[44] Lepor NE, Kereiakes DJ. The PCSK9 inhibitors: A novel therapeutic target enters clinical practice. American Health & Drug Benefits. 2015;**8**(9):483-489

[45] Landmesser U, Chapman MJ, Stock JK, et al. 2017 update of ESC/EAS task force on practical clinical guidance for proprotein convertase subtilisin/kexin type 9 inhibition in patients with atherosclerotic cardiovascular disease or in familial hypercholesterolaemia. European Heart Journal. 2017;**0**:1-13

[46] Sabatine MS, Giugliano RP, Keech AC, et al. FOURIER steering committee and investigators. Evolocumab and clinical outcomes in patients with cardiovascular disease. The New England Journal of Medicine. 2017;**376**:1713-1722

[47] Nissen SE, Stroes E, Dent-Acosta R, et al. Efficacy and tolerability of Evolocumab vs Ezetimibe in patients with muscle-related statin IntoleranceThe GAUSS-3 randomized clinical trial. Journal of the American Medical Association. 2016;**315**(15):1580-1590. DOI: 10.1001/jama.2016.3608

[48] Macedo AF, Taylor FC, Casas JP, et al. Unintended eff ects of statins from observational studies in the general population: Systematic review and meta-analysis. BMC Medicine. 2014;**12**:51

[49] SEARCH Collaborative Group. Intensive lowering of LDL cholesterol with 80 mg versus 20 mg simvastatin daily in 12 064 survivors of myocardial infarction: A double-blind randomised trial. Lancet. 2010;**376**:1658-1669

[50] Ridker PM. The JUPITER Trial-results, controversies, and implications for prevention. Circulation: Cardiovascular Quality and Outcomes. 2009;**2**:279-285

[51] Huisa BN, Stemer AB, Zivin JA. Atorvastatin in stroke: A review of SPARCL and subgroup analysis. Vascular Health and Risk Management. 2010;**6**:229-236

[52] Wu H, Shang H, Wu J. Effect of ezetimibe on glycemic control: A systematic review and meta-analysis of randomized controlled trials. Endocrine. 2018;**60**(2):229-239. DOI: 10.1007/s12020-018-1541-4

[53] Saxon DR, Eckel RH. Statin intolerance: A literature review and management strategies. Progress in Cardiovascular Diseases. 2016;**59**(2):153-164. DOI: 10.1016/j.pcad.2016.07.009

[54] Olsson AG, Angelin B, Assmann G, et al. Can LDL cholesterol be too low? Possible risks of extremely low levels. Journal of Internal Medicine. 2017;**281**:534-553

[55] Khan SU, Talluri S, Riaz H, et al. A Bayesian network meta-analysis of PCSK9 inhibitors, statins and ezetimibe with or without statins for cardiovascular outcomes. European Journal of Preventive Cardiology. 2018;**25**(8):844-853. DOI: 10.1177/2047487318766612

Chapter 7

Designating Vulnerability of Atherosclerotic Plaques

Bukem Tanoren Bilen

Abstract

Microcalcification is an indication of vulnerability of plaques in humans. With conventional imaging modalities, screening of micrometer-sized structures in vivo with high spatial resolution has not been achieved. The goal of this study is to evaluate the potentials of micro-computed tomography (micro-CT), scanning electron microscopy (SEM), energy dispersive X-ray spectroscopy (EDS), time-resolved fluorescence spectroscopy (TRFS), scanning acoustic microscopy (SAM), and photoacoustic microscopy (PAM) in the determination of atherosclerotic plaques with microcalcifications and, therefore, the prospect of constructing a modality on a catheter system. The discrimination of microcalcifications within the fibrocalcific plaques and, therefore, the effectivity of these imaging techniques are discussed. The potential of quantum dots (QDs) in biological imaging is also elucidated since they attract great attention as contrast and therapeutic agents, owing unique properties including good light stability, low toxicity, strong fluorescence intensity, and changing emission wavelength with QD size, ranging from 10 to 100 Å in radius.

Keywords: atherosclerosis, vulnerable plaques, computed tomography, time-resolved fluorescence spectroscopy, scanning electron microscopy, scanning acoustic microscopy, photoacoustic microscopy, microcalcification, quantum dots

1. Introduction

Thin-cap fibroatheromas (TCFAs), which have fibrous caps of thickness of <65 μm [1], are found to be at high risk for rupture. TCFAs have necrotic cores and also calcium depositions [2, 3]. Indicators of plaque vulnerability are microcalcifications or spotty calcifications within the plaques [4, 5], not larger calcifications, which are found to be stable and no longer threatening [6, 7]. Methods such as computed tomography (CT) and echocardiography, which are conventional noninvasive imaging modalities, can detect advanced calcifications. On the other hand, magnetic resonance imaging (MRI), micro-optical coherence tomography (micro-OCT) or positron emission tomography (PET) can identify early calcifications with limitations [6, 8]. However, these modalities are either very expensive or involve ionizing radiation. Therefore, seeking an alternative technique, which can give both morphological and chemical information about tissues at subcellular level, is inevitable.

Micro-CT is an accurate imaging modality for the observation of micrometer-sized structures, which can provide much higher resolution than cone beam computed tomography (CBCT) can. Quantitatively, micro-CT generally has a resolution of less than 10 μm voxel size, while CBCT has a resolution ranging between 76 and

400 μm [9–11]. Using X-rays, micro-CT can perform in vitro processing of the structure of materials such as composites, polymers, and biological materials (bone, teeth, cartilage tissue) and imaging of up to four different substances in a material [12] by a simple sample preparation and positioning procedure and without a requirement of high vacuum or low temperatures that may adversely affect the structure. However, micro-CT is an expensive diagnostic imaging technique requiring an ionizing radiation.

Scanning electron microscopy (SEM) can obtain images with a resolution in the order of a nanometer, by scanning the surface of the specimen with a focused beam of electrons. Energy dispersive X-ray spectroscopy (EDS), which can be implemented in electron microscope systems, is a chemical characterization technique detecting all elements ranging from beryllium (Be) to uranium (U) and their distributions within samples, also by the bombardment of the specimen surface with a focused electron beam. Consequently, both morphological and chemical information about the sample can be obtained by SEM-EDS. Microcalcifications can easily be observed with SEM-EDS; however, this would again be a very expensive method.

Time-resolved fluorescence spectroscopy (TRFS) measures the average fluorescence lifetime that a fluorophore spends in a biological tissue in the excited state, when it is excited by a source from its ground state, and the lifetime of the fluorophore changes as a result of interaction with its molecular environment. Various parameters regarding the molecular environment such as binding, temperature, or concentration can be analyzed by the change in the lifetime. TRFS has not been implemented in clinics yet, however, it has been investigated widely as a new tool for the characterization of atherosclerotic plaques. The success of TRFS in obtaining compositional information about the plaques [13–15] inspired scientists to combine fluorescence lifetime imaging with other modalities such as IVUS [16], second harmonic generation (SHG) microscopy [17], OCT [18], and Raman spectroscopy [19]. Intravascular catheters using TRFS technique have also been built [20, 21], since TRFS catheters can obtain good signals even within the artery, where blood does not affect lifetime properties but fluorescence intensity.

High axial and lateral resolutions of around 20–100 μm, a good penetration depth of around 5 mm, and low cost makes ultrasound imaging very popular for the observation of soft tissue; however, it can only provide morphological information. Besides, the signal detection capability of conventional ultrasonography has to be increased for the detection of microcalcifications, since high echo signals from such small surfaces are not available [22]. These disabilities have been overcome by combining ultrasound with photoacoustic imaging, and the detection of lipid-laden plaque was achieved by providing both morphological and lipid-specific chemical information about the human coronary artery [23]. Photoacoustic microscopy (PAM) is a well-known imaging modality that combines optical and ultrasound imaging. In this technique, typically, nanosecond lasers excite the tissue, and absorbed photons lead to pressure waves via thermoelastic expansion [24]. Ultrasonic transducers capture the emerged pressure waves and produce the map of optical absorbers located within the tissue. Since ultrasonic waves scatter less in biological tissue as opposed to visible portion of electromagnetic spectrum, whole body imaging is possible with a tomographic approach [25]. To increase the penetration even further, lasers operating in the near-infrared region are preferred for excitation where tissue is relatively transparent.

Scanning acoustic microscopy (SAM) is an imaging modality which gives information about the morphology and the mechanical properties of the specimen simultaneously at microscopic levels. High-frequency ultrasound signals are focused to identify the elastic properties of biological tissues. Major advantages of SAM over other imaging techniques are no requirement of special staining and capture of an image of an area of around 5 mm x 5 mm in a couple of minutes. The speed of sound (SOS) through tissues [26–35] or acoustic impedance of samples [36, 37] can be calculated by SAM, and two-dimensional distributions are mapped. Similarly, cells and organelles can be resolved by acoustic microscopy using higher frequencies of 100–1200 MHz [38–45].

Quantum dots (QDs) are used as contrast and therapeutic agents since they have unique properties including strong fluorescence intensity with excellent light stability, low toxicity, and changing emission wavelength with QD size, ranging from 10 to 100 Å in radius [46]. Therefore, they possess great potential in the fields of biological imaging, molecular markers [47–49], and drug delivery [50–52]. QDs are useful in tumor detection [53], cardiovascular imaging [54], and cancer targeting [55]. Their high optical absorption and biocompatibility made noble metal nanoparticles to be widely used as biomarkers [56, 57]. On the other hand, graphene quantum dots (GQDs) are extensively used in biomedical applications [58, 59] since they exhibit lower cytotoxicity than cadmium (Cd), selenium (Se)-, and lead (Pb)-based quantum dots do [60]. The use of QDs as contrast agents in magnetic resonance imaging (MRI) [61], optical coherence tomography (OCT) [54], positron emission tomography (PET), single-photon emission computed tomography (SPECT), optical imaging such as fluorescence spectroscopy and Raman spectroscopy, and photoacoustic imaging (PAI) [62–64] proved their potential as diagnostic agents.

Here, we discuss the ability of these modalities in discriminating the collagen-rich areas from calcified regions within human carotid plaques. Micro-CT and SEM monitors microcalcifications, while EDS provides elemental distribution within plaques. TRFS provides information about the molecular environment of the plaques with the help of QDs. PAM is successful in imaging the fibrocalcific plaques with micrometer resolution. SAM provides micrometer resolution in morphology and also mechanical information about the samples, and therefore, differentiation of the collagen-rich areas from calcified regions is achieved.

2. Imaging of human fibrocalcific plaques

2.1 Micro-computed tomography (micro-CT)

Plaques fixed within 2% formaldehyde can be monitored with micro-CT, after obtaining micro-focal spot and arranging high-resolution detectors for X-rays. With full-scan mode 360° for each plaque, calcifications spread through are observed as can be seen in **Figure 1**.

2.2 Scanning electron microscopy and energy dispersive X-ray spectroscopy (SEM-EDS)

SEM images of the collagen-rich and calcified regions of the plaques can be obtained as shown in **Figure 2** and in **Figure 3**, respectively. EDS analyses of representative regions 2 and 3 are shown in **Figure 4** and in **Figure 5**, respectively. SEM images can be used to differentiate collagen-rich and calcific regions in fibrocalcific plaques, and EDS analyses are performed for determining the compositional differences between two regions of interest. In **Figure 2**, collagen-rich region in one plaque sample is observed and one region is chosen for EDS analysis. Similarly, in **Figure 3**, calcification-rich region is visualized in one sample and one region is chosen for EDS analysis. As can be seen in **Figure 4**, in collagen-rich regions, calcium deposition is insignificant, while in calcified regions, as shown in **Figure 5**, calcium peak is pronounced.

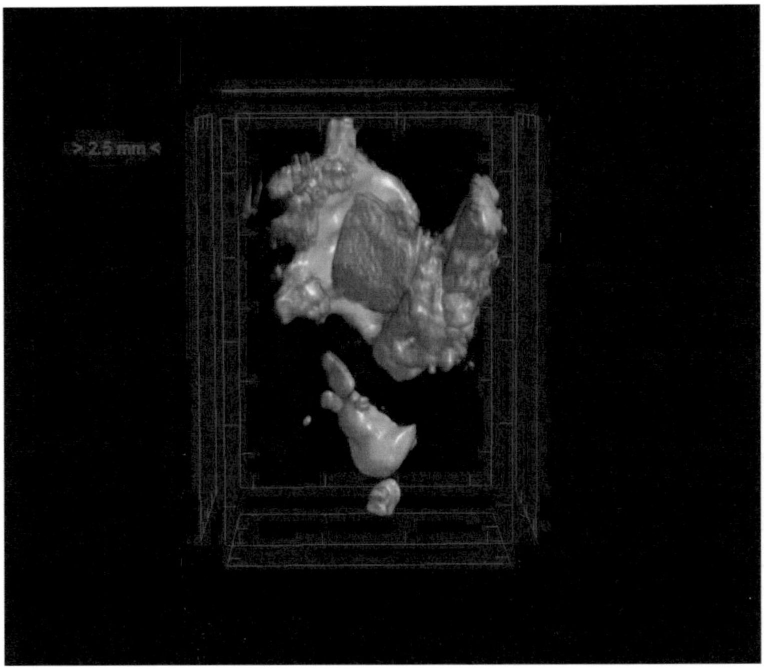

Figure 1.
Micro-CT image of a fibrocalcific plaque sample in which the calcifications can be monitored clearly in three dimensions.

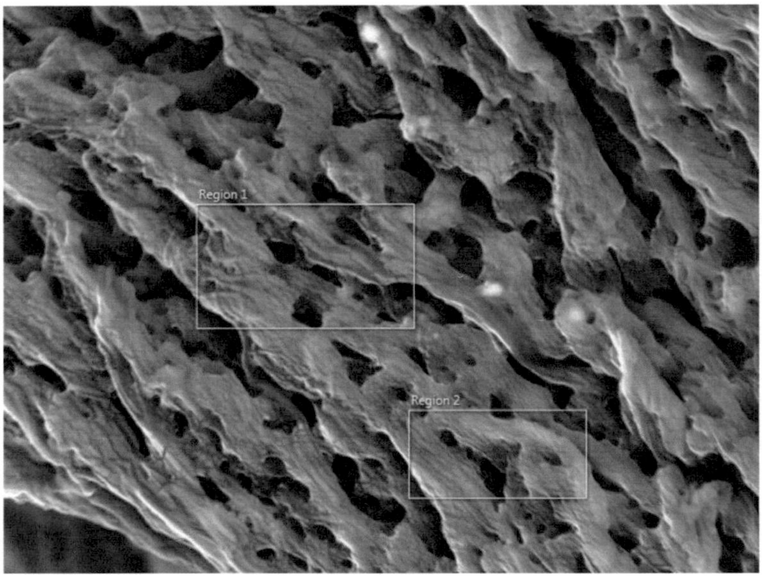

Figure 2.
Scanning electron microscopy image of the collagen-rich region of the plaque. Energy dispersive X-ray spectroscopy is performed on the designated region 2.

2.3 Time-resolved fluorescence spectroscopy (TRFS)

For TRFS experiments, cadmium-telluride/cadmium sulfide (CdTe/CdS) QDs are sprayed on plaque samples and fluorescence lifetimes of the QDs are determined.

The decay of the fluorescence intensity $I(t)$ at time t is given as

$$I(t) = \sum_{i=1}^{n} A_i \exp\left(\frac{-t}{\tau_i}\right) \tag{1}$$

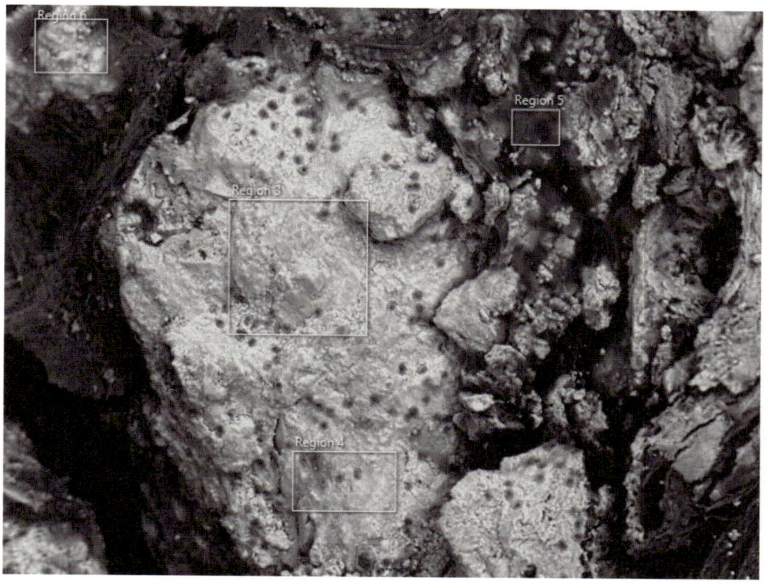

Figure 3.
Scanning electron microscopy image of the calcific region of the plaque. Energy dispersive X-ray spectroscopy is performed on the designated region 3.

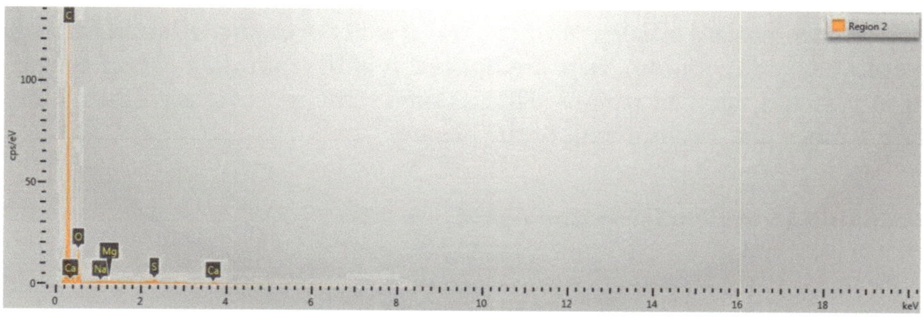

Figure 4.
Energy dispersive X-ray spectroscopy result of the highly calcified region 2, shown in ***Figure 2****.*

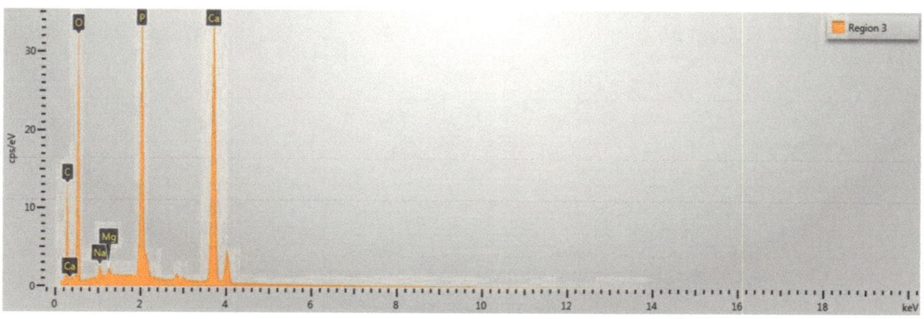

Figure 5.
Energy dispersive X-ray spectroscopy result of the highly calcified region 3, shown in ***Figure 3****.*

where τ_i represents the fluorescence lifetime of the ith component and A_i is its corresponding decay amplitude. The fractional impact of the components to the total intensity is given by

$$f_i = \frac{A_i\tau_i}{\sum_i A_i\tau_i} \quad (2)$$

The intensity decay is evaluated using the average intensity lifetime or the average amplitude lifetime. The amplitude average lifetime is obtained from

$$<\tau> = \sum_i f_i\tau_i \quad (3)$$

The intensity average lifetime is obtained from

$$\tau = \frac{\sum_i A_i\tau_i}{\sum_i A_i} \quad (4)$$

The fluorescence lifetime of QDs is measured on a microscope slide prior to the experiment for the comparison of the lifetime value with those measured on the plaque samples. The lifetime of the QDs on the microscope slide is measured to be 9.24 ns. As seen in **Table 1**, the fluorescence lifetime values of the QDs on a plaque sample are different on various regions.

Figure 6 shows the fluorescence lifetime decay curves acquired from the excitation of the CdTe/CdS QDs on the plaque sample for regions 1, 3, and 4. For the different regions of the sample, the parameters of the decay populations of the QDs confined on the plaque are determined by the two exponential decay fit, minimizing the χ^2 parameter. A significant change is noticeably seen in the characteristics of these three curves for various regions of the sample. These results clearly show that there is an obvious and efficient electron transfer between the QDs and the regions of the plaque, and therefore, there are noticeably different decay parameters at collagen-rich and calcified regions. TRFS is successful in providing information about the molecular environment of the plaque.

2.4 Scanning acoustic microscopy (SAM)

SAM images of the atherosclerotic plaques are received using acoustic impedance mode of SAM as can be seen in **Figure** 7. This image is constructed using the acoustic reflections from both surfaces of the reference (water) and the plaque cross

Region	τ_1 (ns)	τ_2 (ns)	$<\tau>$ (ns)	$<\tau>$ (ns)	χ^2
1	2.90	0.61	1.31	2.17	0.82
2	3.35	0.72	1.36	2.30	0.87
3	4.30	1.00	1.90	3.03	0.86
4	4.73	1.09	2.30	3.60	1.03
5	4.54	1.01	1.91	3.15	1.03
6	4.87	1.06	2.09	3.50	0.93
7	3.70	0.83	1.44	2.41	0.97

Table 1.
Fluorescence decay parameters of CdTe/CdS QDs on different regions of the plaque.

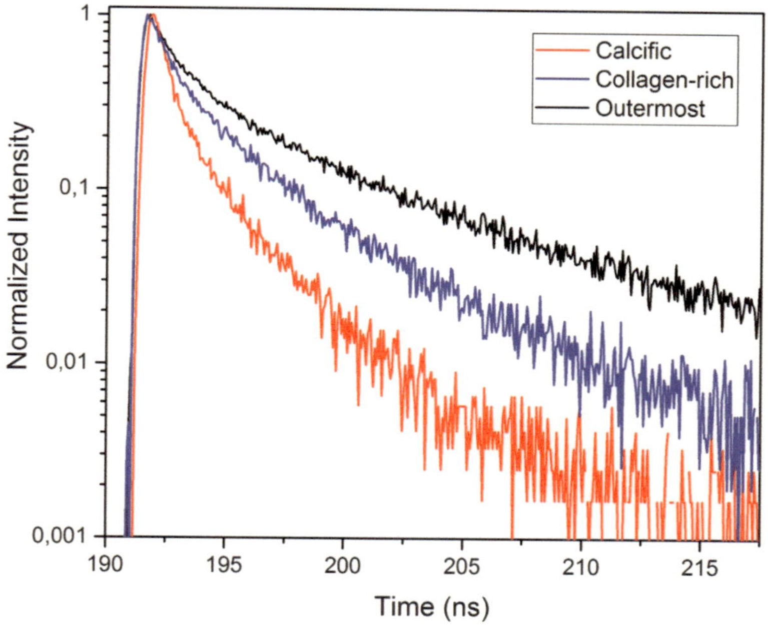

Figure 6.
Fluorescence decay curves of the CdTe/CdS QDs on regions of 1, 3, and 4 on a plaque sample.

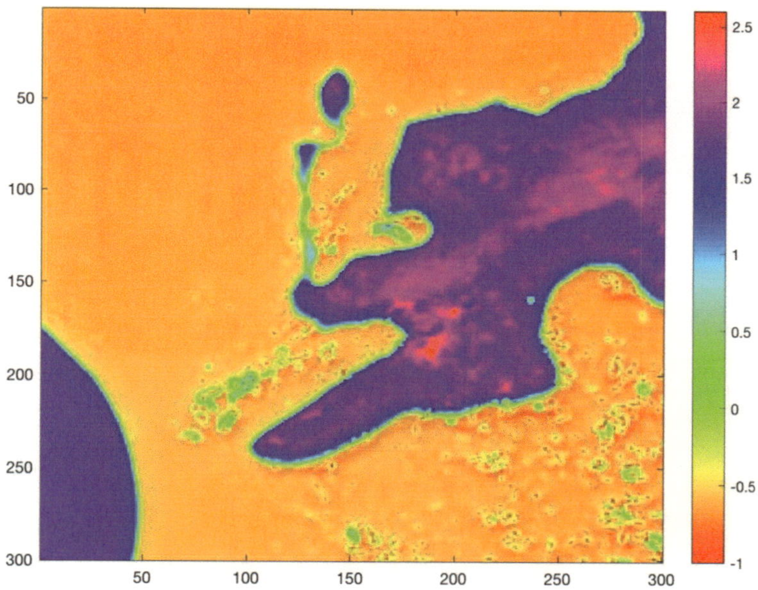

Figure 7.
Acoustic impedance map of a severely calcific plaque sample obtained with acoustic impedance mode of SAM. The scanning area is 4.8 mm x 4.8 mm.

section on the polystyrene substrate. The acoustic impedance distribution indicates different acoustic properties due to the variation of elasticity within the atherosclerotic plaques. The acoustic impedance is determined to be less than 2 MRayl for the collagen-rich areas and greater than 2 MRayl for the calcified areas.

SAM in acoustic impedance mode measures the acoustic impedance of the target by comparing the reflected signal from the tissue with the one from the reference. The reflected signal from the reference is

$$S_{ref} = \frac{Z_{ref} - Z_{sub}}{Z_{ref} + Z_{sub}} S_0 \tag{5}$$

where S_0 is the signal generated by the transducer of SAM, Z_{ref} is the reference's acoustic impedance (1.50 MRayl), and Z_{sub} is the substrate's acoustic impedance (2.37 MRayl). The signal reflected by the target is

$$S_{target} = \frac{Z_{target} - Z_{sub}}{Z_{target} + Z_{sub}} S_0 \tag{6}$$

Consequently, the target's acoustic impedance is calculated as

$$Z_{target} = \frac{1 + \frac{S_{target}}{S_0}}{1 - \frac{S_{target}}{S_0}} Z_{sub} \tag{7}$$

2.5 Photoacoustic microscopy (PAM)

An optically resolved setting (OR-PAM), where focused spot-size on the sample determines the resolution of the system, is successful in imaging the atherosclerotic plaques, as can be seen in **Figure 8**. Calcific regions with greater acoustic impedance values (**Figure 8b**) can also be discriminated by PAM (**Figure 8c**).

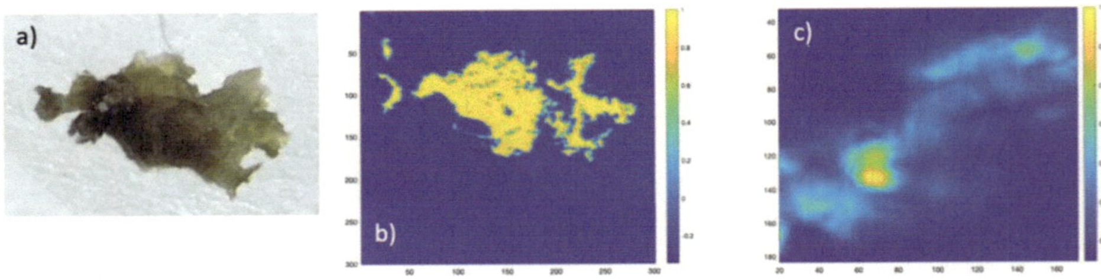

Figure 8.
(a) Digital image, (b) normalized acoustic impedance map, and (c) photoacoustic image of the sample.

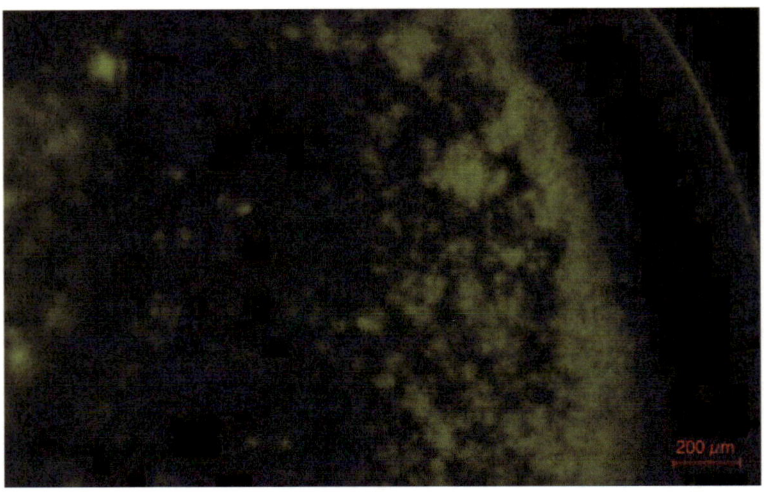

Figure 9.
Photoluminescence image of CdTe/CdS QDs excited with 430 nm. Scale bar is 200 = µm.

3. Quantum dots (QDs)

Photostable QDs are widely used in imaging systems. The photoluminescence image of CdTe/CdS QD aggregates excited with 430 nm is obtained by an inverted fluorescence microscope, as shown in **Figure 9**. The excellent fluorescence intensity with light stability of QDs makes them favorable as diagnostic agents. Use of CdTe/CdS QDs in TRFS experiments of this study reveals their potential in biomedical applications.

4. Conclusion

Here, we discuss the abilities of the imaging modalities on the determination of plaque components of atherosclerotic fibrocalcific plaques. The determination of collagen and calcification within the plaques is done successfully. Micro-CT, SEM, and PAM monitors the microcalcifications. EDS provides elemental distribution within plaques, while TRFS provides information about the molecular environment of the plaques by measuring the lifetime values of CdTe/CdS QDs. SAM provides micrometer resolution in morphology and also mechanical information about the samples. Acoustic impedance maps of the samples show clearly different values in collagen-rich and calcified regions. Consequently, SAM seems predominant over other modalities since SAM is capable of acquiring morphological and chemical information about the plaques simultaneously and usable in clinics. However, for in vivo studies, first, an intravascular SAM probe, similar to intravascular ultrasound (IVUS) probe, has to be developed.

Acknowledgements

Micro-CT experiments were performed in 3D Medical and Industrial Design Laboratory in Istanbul University. SEM-EDS experiments were performed in Nanotechnology Research and Application Center in Sabancı University. TRFS experiments were performed in Photonics Laboratory in Bogazici University. SAM and PAM experiments were performed in Biological and Medical Laboratory in Bogazici University.

Conflict of interest

The author declares no conflict of interest.

Notes/thanks/other declaration

I want to thank all the colleagues who helped me to discuss the characterization techniques mentioned here. Thanks to Leyla Sener Turker for micro-CT experiments. Thanks to Meltem Sezen for SEM-EDS experiments. Thanks to M. Naci Inci for TRFS experiments. Many thanks to M. Burcin Unlu for allowing me to use all of the facilities of Biological and Medical Laboratory not only for SAM and PAM experiments but also for many other research topics.

Appendices and nomenclature

Notation	Definition	Location first used
I	Fluorescence intensity	Eq. (1)
A	Decay amplitude	Eq. (1)
τ	Fluorescence lifetime	Eq. (1)
f	Fractional impact	Eq. (2)
Z_{ref}	Reference's acoustic impedance	Eq. (5)
Z_{sub}	Substrate's acoustic impedance	Eq. (5)
S_{ref}	Reflected signal from the reference	Eq. (5)
S_0	Signal generated by the transducer	Eq. (5)

Author details

Bukem Tanoren Bilen
Department of Physics, Bogazici University, Istanbul, Turkey

*Address all correspondence to: bukem.bilen1@boun.edu.tr

References

[1] Virmani R, Kolodgie FD, Burke AP, Farb A, Schwartz SM. Lessons from sudden coronary death: A comprehensive morphological classification scheme for atherosclerotic lesions. Arteriosclerosis Thrombosis and Vascular Biology. 2000;**20**:1262-1275

[2] Bentzon JF, Otsuka F, Virmani R, Falk E. Mechanisms of plaque formation and rupture. Circulation Research. 2014; **114**:1852-1866

[3] Insull W. The pathology of atherosclerosis: Plaque development and plaque responses to medical treatment. The American Journal of Medicine. 2009;**122**(1A):s3-s14

[4] Bailey G, Meadows J, Morrison AR. Imaging atherosclerotic plaque calcification: Translating biology. The American Journal of Medicine. 2016; **18**(8):51

[5] Vengrenyuk Y, Carlier S, Xanthos S, Cardoso L, Ganatos P, Virmani R, et al. A hypothesis for vulnerable plaque rupture due to stress-induced debonding around cellular microcalcifications in thin fibrous caps. Proceedings of the National Academy of Sciences. 2006;**103**(40):14678-14683

[6] Shanahan CM. Inflammation ushers in calcification. Circulation. 2007;**116**: 12782-12785

[7] Aikawa E, Nahrendorf M, Figueiredo JL, Swirski FK, Shtatland T, Kohler RH, et al. Osteogenesis associates with inflammation in early-stage atherosclerosis evaluated by molecular imaging in vivo. Circulation. 2007;**116**: 12841-12850

[8] Hjortnaes J, New SEP, Aikawa E. Visualizing novel concepts of cardiovascular calcification. Trends in Cardiovascular Medicine. 2013;**23**(3): 71-79

[9] Van Dessel J, Huang Y, Depypere M, Rubira-Bullen I, Maes F, Jacobs R. A comparative evaluation of cone beam CT and micro-CT on trabecular bone structures in the human mandible. Dentomaxillofacial Radiology. 2013;**42**: 20130145

[10] Acar B, Kamburoğlu K, Tatar I, Arıkan V, Çelik H, Yüksel S. Comparison of micro-computerized tomography and cone-beam computerized tomography in the detection of accessory canals in primary molars. Imaging Science in Dentistry. 2015;**45**:205-211

[11] Bauer JS, Link TM, Burghardt A, Henning TD, Mueller D, Majumdar S. Analysis of trabecular bone structure with multidetector spiral computed tomography in a simulated soft-tissue environment. Calcified Tissue International. 2007;**80**:366-373

[12] Rovaris K, Ferreira LM, Sousa TO, Peroni LV, Freitas DQ, Wenzel A, et al. Feasibility of micro-computed tomography to detect and classify proximal caries lesions in vitro. Dental Research Journal. 2018;**15**(2):123-129

[13] Phipps J, Sun Y, Saroufeem R, Hatami N, Fishbein MC, Marcu L. Fluorescence lifetime imaging for the characterization of the biochemical composition of atherosclerotic plaques. Journal of Biomedical Optics. 2011; **16**(9):096018

[14] Phipps JE, Sun Y, Fishbein MC, Marcu L. A fluorescence lifetime imaging classification method to investigate the collagen to lipid ratio in fibrous caps of atherosclerotic plaque. Lasers in Surgery and Medicine. 2012; **44**:564-571

[15] Phipps J, Sun Y, Hatami N, Fishbein MC, Rajaram A, Saroufeem R, et al. Endoscopic fluorescence lifetime

imaging microcopy (FLIM) images of aortic plaque: an automated classification method. In: Proceedings of SPIE; 23–28 January 2010; San Francisco. SPIE. 2010. p. 754839

[16] Marcu L. Fluorescence lifetime in cardiovascular diagnostics. Journal of Biomedical Optics. 2010;**15**(1):011106

[17] Cicchi R, Baria E, Matthaus C, Lange M, Lattermann A, Brehm BR, et al. Non-linear imaging and characterization of atherosclerotic arterial tissue using combined {SHG} and {FLIM} microscopy. Journal of Biophotonics. 2015;**8**(4):347-356

[18] Jo JA, Park J, Pande P, Shrestha S, Serafino MJ, Jimenez JJR, et al. Simultaneous morphological and biochemical endogeneous optical imaging of atherosclerosis. Cardiovascular Imaging. 2015;**16**:910-918

[19] Dochow S, Fatakdawala H, Phipps JE, Ma D, Bocklitz T, Schmitt M, et al. Comparing Raman and fluorescence lifetime spectroscopy from human atherosclerotic lesions using a bimodal probe. Journal of Biophotonics. 2016;**9**:958-966

[20] Fatakdawala H, Gorpas D, Bishop JW, Bec J, Ma D, Southhard JA, et al. Fluorescence lifetime imaging combined with conventional intravascular ultrasound for enhanced assessment of atherosclerotic plaques: An ex vivo study in human coronary arteries. Journal of Cardiovascular Translational Research. 2015;**8**:253-263

[21] Bec J, Ma DM, Yankelevich DR, Liu J, Ferrier WT, Southhard J, et al. Multispectral fluorescence lifetime imaging system for intravascular diagnostics with ultrasound guidance: In vivo validation in swine arteries. Journal of Biophotonics. 2014;**7**(5):281-285

[22] Kamiyama N, Okomura Y, Kakee A, Hashimoto H. Investigation of ultrasound image processing to improve perceptibility of microcalcifications. Journal of Medical Ultrasonics. 2008;**35**: 97-105

[23] Hui J, Cao Y, Zhang Y, Kole A, Wang P, Yu G, et al. Real-time intravascular photoacoustic-ultrasound imaging of lipid-laden plaque in human coronary artery at 16 frames per second. Scientific Reports. 2017;**7**:1417

[24] Xu M, Wang L. Photoacoustic imaging in biomedicine. Review of Scientific Instruments. 2006;**77**:041101

[25] Zhou Y, Yao J, Wang L. Tutorial on photoacoustic tomography. Journal of Biomedical Optics. 2016;**21**: 061007-061007

[26] Miura K, Egawa Y, Moriki T, Mineta H, Harada H, Baba S, et al. Microscopic observation of chemical modification in sections using scanning acoustic microscopy. Pathology International. 2015;**65**(7):355-366

[27] Y Saijo Y, Filho ES, Sasaki H, Yambe T, Tanaka M, Hozumi N, et al. Ultrasonic tissue characterization of atherosclerosis by a speed-of-sound microscanning system. IEEE Transactions on Ultrasonics Ferroelectrics and Frequency Control. 2007;**54**(8):1571-1577

[28] Akhtar R, Cruickshank JK, Zhao X, Derby B, Weber T. A pilot study of scanning acoustic microscopy as a tool for measuring arterial stiffness in aortic biopsies. Artery Research. 2016;**13**:1-5

[29] Y Saijo Y, Hozumi N, Lee C, Nagao M, Kobayashi K, Oakada N, et al. Ultrasonic speed microscopy for imaging of coronary artery. Ultrasonics. 2006;**44**:e51-e55

[30] Miura K, Katoh H. Structural and histochemical alterations in the aortic valves of elderly patients: A comparative study of aortic stenosis, aortic

regurgitation, and normal valves. Biomed Research International. 2016; **2016**:6125204

[31] Brewin MP, Srodon PD, Greenwald SE, Birch MJ. Carotid atherosclerotic plaque characterization by measurement of ultrasound sound speed in vitro at high frequency, 20 MHz. Ultrasonics. 2014;**54**:428-441

[32] Saijo Y, Miyakawa T, Sasaki H, Tanaka M, Nitta S. Acoustic properties of aortic aneurysm obtained with scanning acoustic microscopy. Ultrasonics. 2004;**42**:695-698

[33] Saijo Y, Ohashi T, Sasaki H, Sato M, Jorgensen CS, Nitta S. Application of scanning acoustic microscopy for assessing stress distribution in atherosclerotic plaque. Annals of Biomedical Engineering. 2001;**29**: 1048-1053

[34] Miura K, Nasu H, Yamamoto S. Scanning acoustic microscopy for characterization of neoplastic and inflammatory lesions of lymph nodes. Scientific Reports. 2013;**3**:1255

[35] Miura K, Yamamoto S. Pulmonary imaging with a scanning acoustic microscope discriminates speed-of-sound and shows structural characteristics of disease. Laboratory Investigation. 2012;**92**:1760-1765

[36] Kobayashi K, Yoshida S, Saijo Y, Hozumi N. Acoustic impedance microscopy for biological tissue characterization. Ultrasonics. 2014;**54**: 1922-1928

[37] Hatori K, Saijo Y, Hagiwara Y, Naganuma Y, Igari Kand K, Iikubo M, et al. Acoustic diagnosis device for dentistry. Interface Oral Health Science. 2016;**2016**:181-201

[38] Strohm EM, Czarnota GJ, Kolios MC. Quantitative measurements of apoptotic cell properties using acoustic microscopy. IEEE Transactions on Ultrasonics Ferroelectrics and Frequency Control. 2010;**57**(10):2293-2304

[39] Shelke A, Brand S, Kundu T, Bereiter-Hahn J, Blasé C. Mechanical property quantification of endothelial cells using scanning acoustic microscopy. In: Proceedings of SPIE; 11–15 March 2012; San Diego, SPIE. 2012. p. 83481T

[40] Brand S, Czarnota GJ, Kolios MC, Weiss EC, Lemor R. Visualization of apoptotic cells using scanning acoustic microscopy and high frequency ultrasound. In: Proceedings of the IEEE Ultrasonics Symposium 2; 18-21 September 2005; Rotterdam. IEEE. 2005. pp. 882-885

[41] Masugata H, Mizushige K, Senda S, Kinoshita A, Lu X, Sakamoto H, et al. Tissue characterization of myocardial cells by use of high-frequency acoustic microscopy: Differential myocyte sound speed and its transmural variation in normal, pressure-overload hypertrophic, and amyloid myocardium. Angiology. 1999;**50**(10): 837-845

[42] Miura K, Yamamoto S. A scanning acoustic microscope discriminates cancer cells in fluid. Scientific Reports. 2015;**5**:15243

[43] Strohm EM, Kolios MC. Quantifying the ultrasonic properties of cells during apoptosis using time resolved acoustic microscopy. In: IEEE International Ultrasonics Symposium; 20–23 September 2009; Rome. IEEE. 2009. p. 11208710

[44] Saijo Y, Sasaki H, Sato M, Nitta S, Tanaka M. Visualization of human umbilical vein endothelial cells by acoustic microscopy. Ultrasonics. 2000; **38**:396-399

[45] Soon TTK, Chean TW, Yamada H, Takahashi K, Hozumi N, Kobayashi K,

et al. Effects of anticancer drugs on glia-glioma brain tumor model characterized by acoustic impedance microscopy. Japanese Journal of Applied Physics. 2017;**56**:07JF15

[46] Yoffe AD. Semiconductor quantum dots and related systems: Electronic, optical, luminescence and related properties of low dimensional systems. Advances in Physics. 2001;**50**(1):1-208

[47] Kim H, Beack S, Han S, Shin M, Lee T, Park Y, et al. Multifunctional photonic nanomaterials for diagnostic, therapeutic, and theranostic applications. Advanced Materials. 2018; **30**:1701460

[48] McHugh KJ, Jing L, Behrens AM, Jayawardena SJ, Tang W, Gao M, et al. Biocompatible semiconductor quantum dots as cancer imaging agents. Advanced Materials. 2018;**30**:1706356

[49] Smith AM, Duan H, Mohs AM, Nie S. Bioconjugated quantum dots for in vivomolecular and cellular imaging. Advanced Drug Delivery Review. 2008; **60**:1226-1240

[50] Ohta S, Yamura K, Inasawa S, Yamaguchi Y. Aggregates of silicon quantum dots as a drug carrier: Selective intracellular drug release based on pH-responsive aggregation/dispersion. Chemical Communications. 2015;**51**: 6422

[51] Justin R, Taob K, Román S, Chen D, Xu Y, Geng X, et al. Photoluminescent and superparamagnetic reduced graphene oxide–ironoxide quantum dots for dual-modality imaging, drug delivery and photothermal therapy. Carbon. 2016;**97**:54-70

[52] Zrazhevskiy P, Sena M, Gao X. Designing multifunctional quantum dots for bioimaging, detection, and drug delivery. Chemical Society Reviews. 2010;**39**:4326-4354

[53] Papagiannaros A, Upponi J, Hartner W, Mongayt D, Levchenko T, Torchilin V. Quantum dot loaded immunomicelles for tumor imaging. BMC Medical Imaging. 2010;**10**:22

[54] Hu J, Ortgies DH, Torres RA, Fernandez N, Porto L, Rodriguez EM, et al. Quantum dots emitting in the third biological window as bimodal contrast agents for cardiovascular imaging. Advanced Functional Materials. 2017;**27**: 1703276

[55] Ko NR, Nafiujjaman M, Lee JS, Lim HN, Lee YK, Kwon IK. Graphene quantum dot-based theranostic agents for active targeting of breast cancer. Royal Society of Chemistry Advances. 2017;**7**:11420

[56] Yamazaki R, Ogasawara K, Fujiwara M, Kobayashi K, Saijo Y. Macrophage with gold nanorod visualized by optical-resolution and acoustic-resolution photoacoustic microscopes. In: Conference Proceedings of IEEE Engineering in Medicine and Biology Society; 25–29 August 2015; Milano. IEEE. 2015. pp. 2387-2390

[57] Ha S, Tripathy S, Carson A, Lavery LL, Zhang H, Agarwal A, et al. Multi-target photoacoustic molecular imaging of cardiovascular inflammatory biomarkers using bioconjugated gold nanorods. In: Proceedings of SPIE; 22–27 January 2011; San Francisco. SPIE. 2011. p. 78991M

[58] Thakur M, Kumawat MK, Srivastava R. Multifunctional graphene quantum dots for combined photothermal and photodynamic therapy coupled with cancer cell tracking applications. Royal Society of Chemistry Advances. 2017;**7**:5251

[59] Yang K, Zhang S, Zhang G, Sun X, Lee S, Liu Z. Graphene in mice: Ultrahigh In vivo tumor uptake and

efficient photothermal therapy. Nano Letters. 2010;**10**:3318-3323

[60] Chong Y, Ma Y, Shen H, Tu X, Zhou X, Xu J, et al. The in vitro and in vivo toxicity of graphene quantum dots. Biomaterials. 2014;**35**:5041-5048

[61] Wang H, Revia R, Wang K, Kant RJ, Mu Q, Gai Z, et al. Paramagnetic properties of metal-free boron-doped graphene quantum dots and their application for safe magnetic resonance imaging. Advanced Materials. 2017; **29**(11):1605416

[62] Hong H, Chen F, Cai W. Pharmacokinetic issues of imaging with nanoparticles: Focusing on carbon nanotubes and quantum dots. Molecular Imaging and Biology. 2013;**15**:507-520

[63] Nguyen KT, Sreejith S, Joseph J, He T, Borah P, Guan EY, et al. Poly (acrylic acid)-capped and dye-loaded graphene oxide-mesoporous silica: A nano-sandwich for two-photon and photoacoustic dual-mode imaging. Particle and Particle Systems Characterization. 2014;**31**:1060-1066

[64] Ge J, Jia Q, Liu W, Guo L, Liu Q, Lan M, et al. Red-emissive carbon dots for fluorescent, photoacoustic, and thermal theranostics in living mice. Advanced Materials. 2015;**27**:4169-4177